CONCEPTUAL STATISTICS FOR BEGINNERS

THIRD EDITION

Isadore Newman
Carole Newman
Russell Brown
Sharon McNeely

University Press of America,® Inc.
Lanham · Boulder · New York · Toronto · Oxford

University Press of America,® Inc.
4501 Forbes Boulevard
Suite 200
Lanham, Maryland 20706
UPA Acquisitions Department (301) 459-3366

PO Box 317
Oxford
OX2 9RU, UK

Printed in the United States of America
British Library Cataloging in Publication Information Available

Library of Congress Control Number: 2005935692
ISBN 0-7618-3345-5 (paperback : alk. ppr.)

∞™ The paper used in this publication meets the minimum requirements of American National Standard for Information Sciences—Permanence of Paper for Printed Library Materials, ANSI Z39.48—1984

Contents

Introduction

Most students, when faced with taking a quantitative statistics course, become apprehensive. They are aware that they will be simultaneously confronted with another "language" of terms, concepts, and computations, which they fear will be somewhat overwhelming. Even the most basic statistical textbooks tend to confuse the student.

The purpose of this book is to present the basic concepts, and terminology needed to understand these concepts, using everyday language wherever possible, while also keeping computation at a bare minimum. The emphasis has been almost entirely on developing an understanding of these concepts so that the student can interpret basic quantitative statistics and be better able to critically read the research being done in his or her area of interest.

In some instance, technicalities have been sacrificed for the purpose of conceptual clarity. This was done intentionally, since the authors believe the most appropriate way of learning statistics is first to build an understanding of the underlying principles. It is believed this procedure will provide the student with a firm foundation that will later facilitate the learning of computation and more sophisticated statistics.

Whenever possible, everyday language has been used instead of technical jargon. This is in keeping with research derived from Piaget's theory that indicates that the introduction of new material is most efficiently learned when that material is presented using a language most familiar to the student. Therefore, statistical and mathematical symbolism has been kept to a minimum.

All students who are statistically unsophisticated should be able to find some benefit from reading this monograph. Chapters I, II, and the beginning of Chapter III, which deals with the meaning of statistical significance, should be read by all. These chapters provide background knowledge necessary for successfully reading other chapters. The remainder of Chapter III, and Chapters IV and V introduce and define terms and concepts that are slightly more sophisticated. They will be most beneficial for the student who desires to learn more statistics; also, they will help all students conceptualize the relationship between certain statistical procedures and principles.

Chapter VI is a brief and conceptual introduction to quantitative research design and the concept of validity of the design, based on a tradition started by Campbell and Stanley. It presents a conceptual model on which research designs can be placed on a continuum based on their internal validity. It also discusses external validity. It is a crucial chapter for everyone who is interested in a basic understanding of quantitative research.

Chapter VII is a continuation of concepts presented in Chapter VI and extends them in the Campbell and Stanley tradition. It introduces some designs for conducting research, including quasi-experimental designs and time-series designs.

Chapter VIII is a paper entitled Type VI error that was presented at the World Population Society. It deals with the conceptual problems in which many researchers find themselves entangled. Many sections of this paper are at a more sophisticated level than the rest of the text; however, the first seven pages can be read, understood, and useful to the beginning student.

To facilitate the achievement of the objectives in these chapters, see the hypertext links that are embodied in the text.

Objectives for Chapter I

After completing Chapter I, you should be able to:

1. Define the terms descriptive and inferential statistics.
2. Define the three measures of central tendency and state when it is most appropriate to use each one.
3. Calculate a mean score, median and mode for a distribution
4. Identify the characteristics of a normal distribution.
5. Define a skewed distribution.
6. Identify the relative positions of the mean, median and mode on both a positively and negatively skewed distribution.
7. Define variability.
8. Define range, quartile deviation, and standard deviation and state when it is most appropriate to use each one.
9. Calculate standard deviation.
10. Identify and state the characteristics of the normal curve.
11. Define Z scores, T scores, and stanines and explain how to interpret each one.
12. Calculate a Z score.
13. Define and interpret percentile, percentile rank, age equivalence, and grade equivalence and state the advantages and disadvantages of each one.
14. See hyperlinks.

Chapter I

Introduction to Statistics

This chapter is written for the primary purpose of introducing statistical terminology and concepts. An understanding of the material is basic to the understanding and interpretation of most social science research articles, and to the appropriate use and interpretation of standardized test information.

Statistical procedures can be divided into two types: *descriptive* and *inferential*. When someone has information on a group of people (called a sample of a population), it becomes necessary to find an accurate manner to compile this data for the sake of efficiency and understanding. The statistical procedures developed for this purpose are called descriptive. Descriptive statistics are only used for describing the population on which one has data. They are not to infer to or describe groups on which information has not been collected. Almost all the statistical techniques presented in this chapter will be descriptive. If they are not descriptive, that will be noted.

Inferential statistics, on the other hand, are based on data collected from small samples (subgroups) from a larger group. The sample information, once collected and analyzed, is then used to infer to the entire group (population) from which the sample came. In other words, one makes *predictions* about the entire group (population) based on the information gathered about the sample. The statistical techniques that allow one to predict what a group (population) will do on the basis of information gathered about a sample are called either inferential or sampling statistics. Throughout this text, we will refer to them as inferential statistics. When a researcher is using descriptive or inferential statistical procedures, the group of people from whom the data are collected is called a sample.

Subjects

Our sample of a population will be called our group. The members of that group are called our subjects. Many times group and subjects are used synonymously. We will use "S" to refer to subjects who are from our sample population. We will subnumber the S_1 when we have various groups of subjects from our population.

Subjects are obtained in various ways. Some researchers do not care if the subjects they choose are representative of a larger population. In that case, they use what is called nonprobabilistic sampling. There are three kinds of nonprobabilistic sampling: incidental, quota, and purposive.

Incidental sampling generally refers to picking a sample of subjects because it is convenient. This is the least accurate sampling technique, but it is also the most frequently used.

Quota sampling involves selecting subjects based on some known information. Here, specific information about the proportions that occur within a population is used, and subjects are selected to fit those proportions. For instance, if you were studying nutrition in a population, and you wanted to compare nutrition based on income level, you might select subjects proportional to the known incomes. Here, subjects are picked on a first-come, first use basis, not based on random selection.

Purposive sampling chooses the subjects on the basis of some special purpose. For example, sometimes a political pollster will select people to poll based on their political party affiliation.

If researchers care about the subjects they choose being representative of a larger population, that is, that the sampling is not biased, then they engage in probabilistic sampling. Probabilistic sampling requires that some sort of random sampling is used so that for each subject there is a known probability of being picked as part of a sample from that population. There are four methods of random sampling, all which are modifications of simple random sampling. These are: simple random sampling, systematic sampling, stratified random sampling, and cluster sampling. Random sampling involves the use of a Table of

Random Numbers to pick a random sample. Most Random Number Tables consist of a series of five-digit numbers such as:

35956	27348
95847	63121
73603	59430
19283	43324
84720	12942
32732	04231
09439	75543

Assume that as a researcher you have a population of 500 people from which you want to select a sample. You assign a number to every person in the population. You know you want 30 subjects. Then, arbitrarily you select a column in the table, here, the second column. You are only selecting from a population of 500, so the last three digits of the five are used to make your selection.

- The first of your 30 subjects would be number 348.
- The second would be number 121.
- The third would be number 430.
- The fourth would be number 324.
- The number 942 is not in your group, so it is skipped, making the fifth person number 231.

Simple random sampling is a procedure in which a sample of a population is drawn in such a way that each person has an equal chance of being selected. The sampling described above is considered to be simple random sampling.

Systematic sampling draws only on the first number randomly. Then, each of the other subjects are drawn according to some predetermined plan, such as every tenth person. This sampling technique is usually used when there is a large sample to be drawn as it saves time.

Stratified random sampling is used when some specific strata or variables are important. For instance, the gender, age, and economic status of the population is known. The percentages of these as they occur in the population is maintained when doing the sampling by dividing the population into the subgroups and sampling within each of them.

Cluster sampling breaks the population down into defined groups, called clusters. Once the clusters are defined, a sample cluster is randomly selected from within each cluster. Many times cluster and stratified sampling is used in combination to develop a representative sampling of a population. There is one major concern when cluster sampling is used, that traditional statistical procedures cannot be used.

Measures of Central Tendency: Finding the "Average"

To understand the characteristics of a population, one of the things that statisticians usually consider is describing the samples of the population in terms of the average score. This "average" score is called the central tendency.

There are three basic measures of central tendency: the mean ($\bar{x}$), median (mdn) and mode. Each of these measures produces one score that describes or characterizes the performance of the sample. By using the same measure of central tendency to report the average performance of two or more samples/groups, one can more easily compare the average relative performance of these samples/groups. We will usually have one or two groups that we will use as examples.

Mean Score ($\bar{x}$)

The most often used of the three measures of central tendency is the mean score. Many people frequently use the terms mean and average interchangeably thinking that the mean score is the only type of average one can calculate. While this is the most popular average used, it is not the only one.

The formula for calculating the mean score of a sample/group is:

$$\bar{x} = \frac{\sum x}{N}$$

Where: $\bar{x}$ = mean score
Σx = the sum (Σ) of scores (x) arrived at by adding all the scores together
N = the number of subjects or scores being added

For example, if one is interested in calculating the average mean performance of a group of eleven students on a particular test, one would proceed as follows:

Group One:

Students	*Scores*	*Students*	*Scores*	*Students*	*Scores*
Sidney	95	Silvia	35	Isdaore	65
David	80	Dora	60	Fran	60
Matthew	75	Maria	85	Ivette	25
Carlos	90	Carole	100		
Total (Σx)	770				

In this case, 770 is the sum of all the scores ($\Sigma x = 770$). There were eleven scores added to arrive at the total (N = 11).

$$\bar{x} = \frac{\Sigma x = 770}{N = 11} = 70$$

The single mean score that represents the average performance of this group is 70.

Vassarstats provides the following web page that will calculate the mean for a given set of data:

http://faculty.vassar.edu/ lowry/basic.html.

Median (mdn)

The median is simply defined as the middle point in a range of scores that have been put in order from lowest to highest or highest to lowest (rank ordered). This is used somewhat less than the mean. The median is calculated in such a way that it always lies at the center of a set of scores. This is not necessarily true for a mean score.

For example, we can use the same scores in the previous illustration but we must first rank order them. The scores would then be listed as: 100, 95, 90, 85, 80, 75, 65, 60, 60, 35, and 25. The score of 75 is the median since an equal number of the scores fall above and below it. Notice that all scores are listed, even repeating ones.

It becomes slightly more difficult to calculate the median when there is an even number of scores, because the median will fall between the middle two scores. Assume we remove Ivette's score from the sample above. Now we have the rank scores of 100, 95, 90, 85, 80, 75, 65, 60, 60, and 35. The two scores that fall in the middle of this distribution are 80 and 75. The median will fall evenly between these scores, and in this case is 77.5. Yes, the median score is 77.5, even though no one actually had this score. It was arrived at by locating the two middle scores (80, 75), adding them together (80 + 75 = 155), and then dividing by two (155/2 = 77.5). This median score would then be a point in the distribution where 50% of the scores fall above it and 50% of the scores fall below it.

Just for comparison, look at how the mean score also changed without Ivette in the group. Now the mean score is 745/10 = 74.5.

Mode

The least frequently used estimate of central tendency is the mode. One reason that it is typically not used is that it is likely to change very easily and drastically from sample to sample. It is also likely to be the poorest of the three measures of central tendency for representing the average group performance.

The mode is defined as the most frequently occurring score in a distribution of scores. For example, in the previously defined distribution: 25, 35, 60, 60, 65, 75, 80, 85, 90, 95, 100, the mode is 60 because it is the most frequently occurring score in the distribution.

If we had a distribution in which two sets of scores occurred with equal frequency, each would be a mode. The distribution would be called bimodal. Assume we added one more score of 90 to the sample. Now we would have the scores of: 25, 35, 60, 60, 65, 75, 80, 85, 90, 90, 95, 100. Both 60 and 90 would both be modes of this distribution. If a distribution occurred in which three scores appeared with equal frequency, it would be called trimodal.

Appropriate Use of the Measures of Central Tendency

To determine which measure of central tendency to use, typically statisticians look at the distribution of the scores. This is done by plotting all of the scores onto a diagram, and then drawing the shape of this plot. The shape of the distribution is then used to determine which measure of central tendency to use. In some situations, the distribution will be labeled as normal, and all three measures of central tendency will produce the same estimate. In other situations, the distribution will be labeled as skewed, and the estimate will be drastically different. In the next few sections, we will explore this idea further.

What is data?

When we have a set of scores, we call the scores data. When we have one score, it is called datum.

Normal Distribution of Data

We assume data is normally distributed when the most frequently occurring scores are grouped in the middle of the distribution: and as scores move away from the middle in both directions (higher or lower than the average), they steadily decrease in frequency to an equal degree. Normally distributed data will always be in a bell-shaped distribution. Figure 1 is an ex(

Figure 1: Example of a bell shaped distribution

% of scores occurring in the distribution

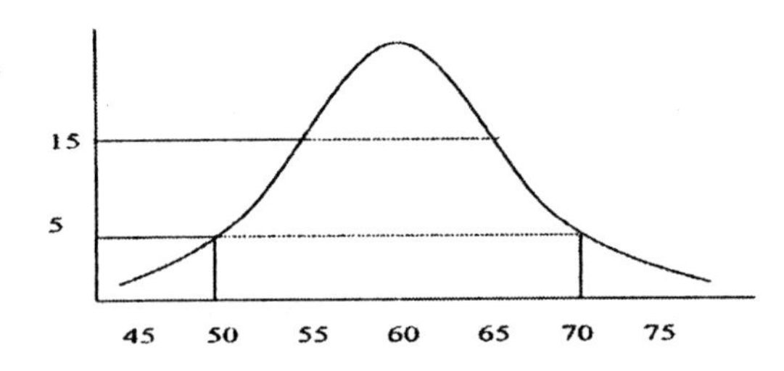

In this Figure, there were approximately an equal number of scores at 55 and 65 and, also an equal but fewer number of scores at 50 and 70.

Very rarely does one's data or scores fall exactly in a normal or bell-shaped distribution. However, if the distribution of scores does not radically differ from this type of distribution, we tend to assume that the scores actually are normal. If the scores are normally distributed, then the mean, median, and mode will fall exactly at the same point. This is one of the characteristics of the normal curve, as is illustrated in Figure 2.

Figure 2: Example of a normal distribution

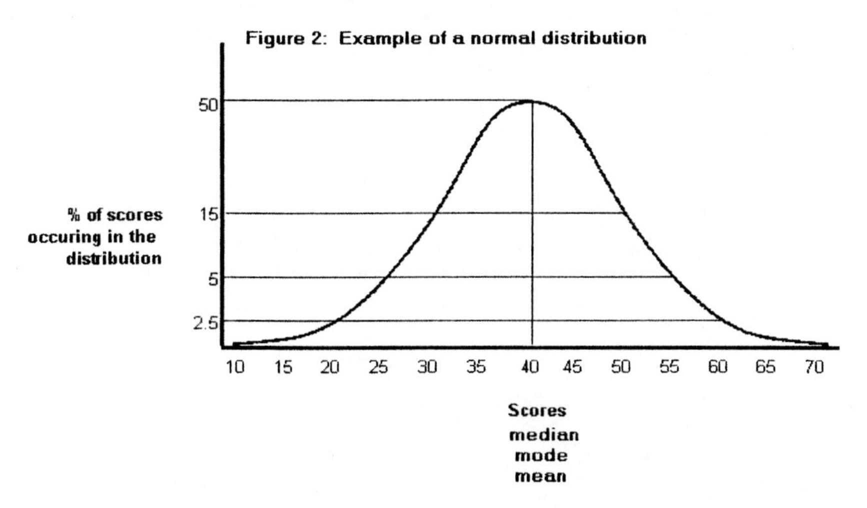

As we can see by looking at Figure 2, the mode is the point in the distribution that occurs most frequently. The median is the point in the distribution where 50% of the scores fall above it and 50% fall below it. The mean score is arrived at by adding all scores in a distribution and then dividing by the number of scores. In a normal distribution, the mean score will be the score that is in the center of all scores. If the distribution is not normal, this is not true.

To conceptually understand the function of the mean score, one may think of a yardstick as being the base of our normal curve with each inch of the stick representing a score. Every time a score is received, we would place one poker chip at that point on the yardstick. For example, if three people received a score represented by the two-inch mark, then three poker chips would be placed on that mark on the yardstick. If this was done for every single score obtained and the scores were normally distributed across all 36 inches, then most of the chips would be piled on the 18-inch mark and numbers on both sides of the 18-inch mark would receive equal but increasingly less chips as we moved away from the score represented by the 18 inches.

If we then took an ice pick and tried to balance the yardstick, the point where the yardstick would be perfectly balanced would be the mean. This would be true for normally and not normally distributed data, but for a normal distribution this balance point will be exactly in the center of all of the scores in that distribution.

When the scores are exactly normally distributed (which is very rare if ever) the mean, median and mode will produce the same estimate of central tendency. However, the mean is the most stable estimate and will tend to fluctuate less. It is also a statistic that is the basis of the calculations of many other statistics. For these reasons, the mean is used more frequently than the other two measures of central tendency, and it is sometimes used inappropriately as will be discussed in the next section.

Skewed Distributions of Data

A skewed distribution is one in which most of the scores are closer to one end than the other. Unlike a normal or symmetrical distribution, a skewed distribution will have most of its scores on one side and, therefore, would not be symmetrical.

Skewed distributions are referred to as being either positively or negatively skewed. In a positively skewed distribution, most of the scores will be on the lower end of the distribution (more low scores than high scores), and the long tail will be pointing toward the higher end of the distribution. To identify if a distribution is positive or negative, put an arrowhead on the tail. If the arrow is pointing to the low end (left), the distribution is negatively skewed. See Figures 3 and 4 that demonstrate the relationship between the mean, median, and mode on skewed distributions.

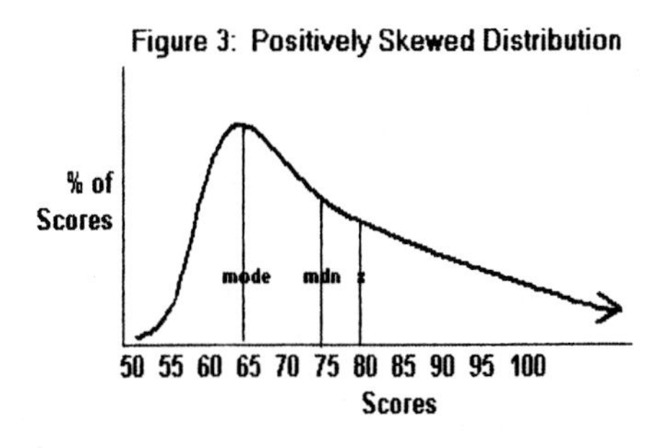

Figure 3: Positively Skewed Distribution

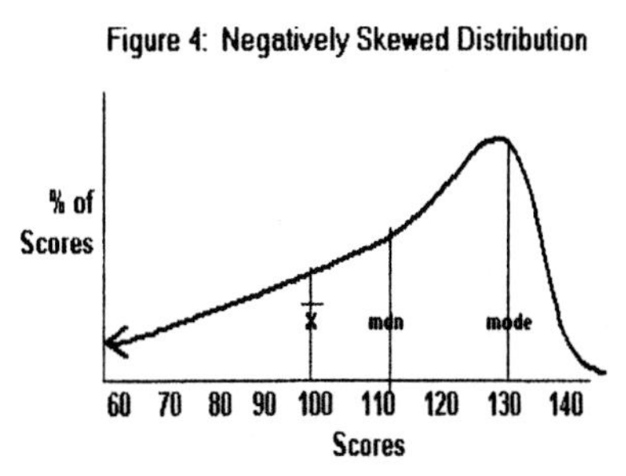

Figure 4: Negatively Skewed Distribution

An examination of Figures 3 and 4 illustrates that the mean in a skewed distribution is always pulled toward the tail. In a positively skewed distribution, the mean will always be higher than the median or mode; and in a negatively skewed distribution, the mean will be lower than the median or mode. The mean is not at the center of the distribution and is affected by extreme scores. The median is not affected by extreme scores. It is the point in the distribution where 50% of the scores fall on either side of it. The mode is the most frequently occurring score, in any distribution.

When scores are skewed, the most stable and representative measure of central tendency would be the median. Let's go back to our earlier sample of 11 scores, and assume these are points earned on homework. Remember, these scores are 100, 95, 90, 85, 80, 75, 65, 60, 60, 35, 25.

If we were interested in determining the average points earned on homework, this number should be representative of most of the subjects. If we used the mean score, we might be led to believe that most of the subjects earned 70 points, which is not representative of the actual points earned. The mean score is affected by extreme scores. The median, on the other hand, is not affected by extreme scores. It is the point in the distribution where 50% of the scores fall above and 50% fall below. The median for the above distribution is 75, which is much more representative of most students' points earned.

If you were asked to calculate the average income in the United States, and if a mean was used, you would determine that the average income would be around $60,000. This is because the distribution is positively skewed and the mean is affected by and pulled toward the extreme. See Figure 5 for an appreciation of this distribution.

Figure 5: Example of a distribution where the mean score is pulled to the extreme

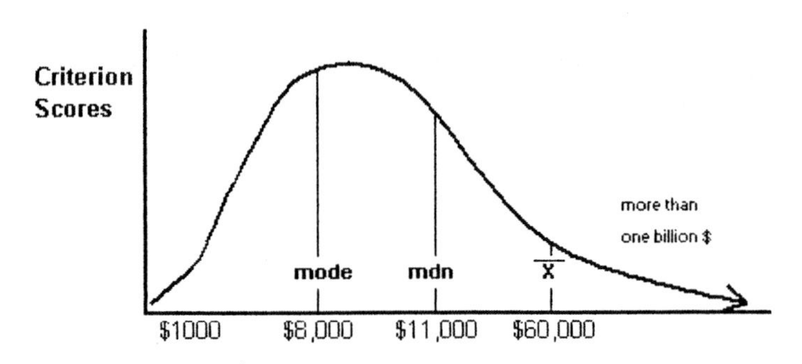

The inappropriate use of statistics, as the mean score is in this case, has led some people to make the charge that "statistics lie." This is an inaccurate statement since the problem generally occurs because of the use of an inappropriate statistic. The mean score should be used only if the data is approximately normally distributed. Use under any other conditions produces a distorted representation. The median can be used for both normal and skewed distributions, but it is only preferable when the data is skewed. This mode is the least accurate and is generally not used. The only possible use it may have in the educational setting is for a quick and rough estimate.

Variability

Variability is defined as the degree the scores differ (vary) around the measure of central tendency. The more representative the measure of central tendency is for all scores, the less variability. The less representatives the central tendency measure, the more the variability.

There are three general estimates of variability: range, quartile deviation, and standard deviation. Of the three, standard deviation is by far the most widely used in quantitative research, and will receive the greatest attention in our discussion.

Range

Range is the simplest and least accurate measure of variability. It is simply defined as the highest score in the distribution minus the lowest score (High – Low). In some texts, it is defined by a two-step process, 1) highest score minus the lowest score, and 2) add one (High – Low + 1). If all the scores in the distribution are very similar, then subtracting the lowest score from the highest will produce a very small difference. This would indicate that there is little variability. At the very best, range is a quick and inaccurate estimate of the variability among scores in a distribution.

For our group 1, the range would be calculated by taking the highest score (100) minus the lowest score (25). This would result in a range score of 75.

Quartile Deviation (Q)

The quartile deviation estimates variability. To calculate this, we have to do two steps. First, we determine the 25% (Q1) and the 75% (Q3) points in a distribution of scores. Q1 is the point in the distribution of scores where 75% of the scores fall above and 25% fall below. Q3 is the point in the distribution where 25% of the scores are above and 75% are below. The median can be called Q2 since it is the point in the distribution where 50% of the scores fall above it and 50% are below it. Second, the actual quartile deviation (Q) is determined by subtracting Q1 from Q3, and then dividing the resulting difference score by two [Q = (Q3 - Q1)/2].

For our group 1, the scores are put in rank order: 100, 95, 90, 85, 80, 75, 65, 60, 60, 35, 25. We have already established that the median (Q2) is 75. Q1 would be 60. Q3 would be 90. The quartile deviation (Q) would be (90-60)/2 = 30/2 = 15.

Quartile deviation should be used whenever it is appropriate to use the median. Fifty percent of all scores in the distribution will fall between the median plus and minus one quartile (mdn ± 1Q).

Variance and Standard Deviation (sd)

The most frequently used estimate of variability is the standard deviation. Conceptually, it is the average amount each individual score differs from the mean score of its group. One way to calculate the standard deviation is to subtract the mean score ($\bar{x}$) from each score and square the differences to eliminate negative numbers. The resulting squares are then summed. This total, the sum of the squared deviations from the mean, is also known as the Sum of Squares (SS). The variance is obtained by dividing the Sum of Squares (SS) by the number of subjects in the population.

$$\text{Variance} = \frac{\sum x^2}{N} = \frac{SS}{N}$$

Where: x^2 = score minus mean score quantity squared
$x - \bar{x}^2$ = deviation score squared
N = number of people in the group*
$\sum x^2$ = sum of all deviation scores squared

The standard deviation is simply the square root of the variance:

$$\sqrt{\frac{\sum x^2}{N}}$$

If we compare the formula for standard deviation to the formula for the mean score ($\bar{x} = \sum x/N$), we can see that conceptually the standard deviation is the average of the squared deviation scores. Using the following scores (group 2 = 1, 2, 3, 4, 5, 6, and 7), we will demonstrate the computation.

X	$(X - \bar{x}) = x$	x^2
1	(1 – 4) = -3	9
2	(2 – 4) = -2	4
3	(3 – 4) = -1	1
4	(4 – 4) = 0	0

	5	(5 – 4) = 1	1
	6	(6 – 4) = 2	4
	7	(7 – 4) = 3	9
Σ	28	0	28

The sum of X = 28.
The mean score is 28/7 = 4
The sum of x = 0
The sum of x^2 = 28.
The mean squared deviation ($\Sigma x^2/N$) = 28/7 = 4
The standard deviation is the square root of 28/7 = the square root of 4 = 2.

For group 1, the mean score is 4, and the standard deviation is 2. In inferential statistics, N-1 rather than N is used in the denominator when dealing with samples to correct for sampling error; therefore, if the scores were a sample from a larger group, the standard deviation would be equal to the square root of 28/6 which equals 2.16.

For further information and additional examples regarding measures of central tendency and variability, we encourage you to visit the following link:

http://psychology.wadsworth.com/workshops/ central1.html.

In turn, the following web page will provide an on-line calculation of the standard deviation (and mean) for a given set of data.:

http://faculty.vassar.edu/lowry/basic.html.

The mean and standard deviation are most useful when the scores are distributed normally. For that, we need to know more about the normal curve.

Normal Distribution

Some of the characteristics of the normal curve are:

1. It is bell shaped.
2. The maximum height is at the mean score
3. It is asymptotic to the X-axis (theoretically, the tails of the curve approach but never touch the X-axis)

4. Approximately 68% of all the cases will fall between the mean score ± one standard deviation (0± 1 sd = 68.26%)
5. 0 ± 2sd = 95.44% (approximately 95% of all cases will fall between the mean score ± two standard deviations)
6. 0 ± 3sd = 99.98% (approximately 99% of all cases will fall between the mean score ± three standard deviations)

In Figure 6, scores are assumed to be normally distributed, with the mean score = 100 and the standard deviation = 15. Given the scores applied to the X-axis, approximately 95% of all scores will fall between 70 and 130; 99% of the scores will fall between 55 and 145. Using the mean score and standard deviation, one can determine the percentage of people who will fall between a particular range of scores. For additional information regarding normal distributions, please see the following web page:

http://davidmlane.com/hyperstat/ normal distribution.html

Figure 6: Percentages of a sample within a normal distribution

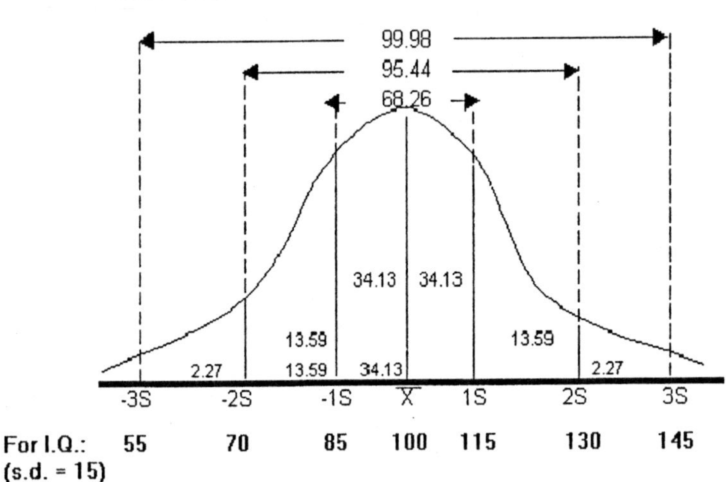

Standard Scores

Frequently, we want, or find it necessary, to compare scores that may have different mean scores and/or different standard deviations. Comparison is facilitated by changing all scores involved to standard scores that will have a common mean score and common standard deviation. The three most frequently used standard scores are Z scores, T scores, and stanines.

The major advantage of standard scores over raw scores is that they have specific mean scores and standard deviations that are always whole numbers and are, therefore, easier to manipulate. They facilitate our ability to differentially weigh any number of scores based on what we perceive to be their importance, and they make it much easier to compare scores in a meaningful and appropriate manner.

A major disadvantage of standard scores is that they assume a normal distribution. If a normal distribution does not exist, it is very difficult to accurately interpret standard scores. There is also a general lack of familiarity on the part of the general public with the terms and use of standard scores. This limits their use since it becomes difficult to convey meaningful information.

Z Scores

Z scores convert the raw scores into standard deviation units. For each distribution, Z always has a mean score of "O" and a standard deviation of 1 ($\bar{x}$ = 0, sd = 1). The formula for Z scores is $Z = (X - \bar{x})/sd = x/sd$

Where: X = any score in a set of score
$\bar{x}$ = the mean of the set of scores
sd = the standard deviation of the set of scores
$x = X - \bar{x}$

Using the data from group 2, the following illustrates how raw scores can be converted to Z scores.

Z Scores Table

X	(X - 0) = x	Z = x/sd
1	(1– 4) = -3	-3/2 = -1.5
2	(2 – 4) = -2	-2/2 = -1.0
3	(3 – 4) = -1	-1/2 = -0.5
4	(4 – 4) = 0	0/2 = 0.0
5	(5 – 4) = 1	1/2 = +0.5
6	(6 – 4) = 2	2/2 = +1.0
7	(7 – 4) = 3	3/2 = +1.5

The sum of x = 28.
The mean score is 28/7 = 4
The sum of x = 0
The sum of x^2 = 28.
The standard deviation is the square root of 28/7
= the square root of 4 = 2.

In the preceding example, the first person received a score of 1. This was converted to a Z = -1.5. This can be interpreted as a score of 1is 1.5 standard deviations below the mean score of the group (4) since Z scores are standard deviation units. Similarly, the sixth person's score of 6 was converted to a Z = +1.0. This individual was 1.0 standard deviation units above the mean score of the group (4).

Why would someone use Z scores? One reason is that they far better represent the data than using raw scores. Consider the case of a teacher who is trying to determine grades for the students. In Table 1 below, we'll look at two students' scores on four different quizzes.

Table 1: Example of two student's scores on four different quizzes

Test	$\bar{x}$	sd	Matt's Test scores	Carole's Test scores	Matt's Z score	Carole's Z score
1	150	50	250	200	2.0	1.0
2	75	25	62.5	25	.5	-2.0

3	50	10	80	50	3.0	0.0
4	300	100	300	500	0.0	2.0

	Matt	*Carole*
Mean scores:	173.1	193.75
Sum of Z scores:	4.5	1.0
Mean of Z scores:	1.125	.25

By just calculating the mean scores, it would appear that Carole did better than Matt. However, this is incorrect since it does not take into consideration the different mean scores and standard deviations. An inspection of the table shows that, on most of the tests, Matt's scores are better than Carole's scores in relationship to the mean scores and standard deviations of each test. Matt's average Z score (Z) on all four tests is 1.125, which means he is 1.125 standard deviations above the mean score while Carole's Z = .25 which is .25 standard deviations above the mean score for all four tests. Changing raw scores to Z scores allows us to compare each person's score to the average performance of the group.

We have shown that a more accurate and correct method of comparing would be to change all of the raw scores to Z scores and compare the Z scores since the Z scores more accurately show each person's relative position in the group; that is, where each score fits in relation to the mean of the group.

Looking at the table again, we see that on Test 3, Carole has a Z = 0. This indicates that her score on that test was exactly at the mean of her group for that test. On Test 2, Matt's Z = -.5 indicates that he was 1/2 a standard deviation below the average of his group for Test 2.

We can also interpret Z scores in terms of the percentage of group members scoring above or below a particular Z score.

On Test 1, Matt's Z = 2. This means he scored two standard deviation units above the average on Test 1. By looking at Figure 7 one can determine the percentage of people in the group who scored lower than Matt on that particular test. In this case, approximately 97% of the people scored lower than Matt. We can determine this by adding up all the percentages under the curve that are less than 2 standard deviations or 2 Z scores. The following link will calculate the proportion of scores that fall above and below a given Z score: http://faculty.vassar.edu/lowry/zp.html.

Figure 7: Z scores as they relate to the percentages of a sample within a normal distribution

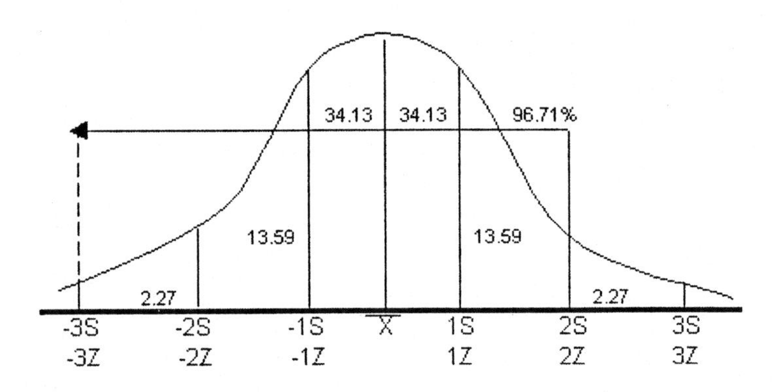

There are many advantages for using Z scores over raw scores. In addition to the ones already discussed, they facilitate the weighting of tests. If we wanted to count one test twice as much as some other test, it is easy to change the raw scores to a Z score and multiply by two. By changing all raw scores to Z scores, it is generally more convenient for recording and scoring since they are always dealing with the same units that are generally within the range of +4 to –4.

One major limitation for using Z scores is that they are a function of the mean score and standard deviation and are subject to the same restrictions. They are only appropriate when the data are normally distributed or approximately normally distributed. For further examples and explanation of z-scores, visit the following link to a series of pages devoted to the subject: http://psychology.wadsworth.com/workshops/zscores1.html.

T Scores

The T score is based on the same concept as the Z score. For any particular distribution of Z scores, the mean score is 0 and the standard deviation is 1, (Z: $\bar{x} = 0$, sd = 1) while for T scores, the mean score is 50 with a standard deviation of 10, (T: $\bar{x} = 50$, sd = 10).

The formula for T scores is T = 50 + 10(Z). First, the Z score is found, multiplied by 10, and then 50 is added to it. This procedure eliminates the negative numbers in working with Z scores. This decreases the likelihood of making clerical errors. This same idea is applied to the College Board Exams on which 500 is added to every score and Z is multiplied by 100.

T scores, like Zs, are subject to the same limitations. If the distribution of scores is not normal, it is difficult to interpret the Ts since they, like Zs, are based on the underlying assumption of normality. Standard scores such as Zs and Ts will have the same distribution as the raw scores. If the raw scores are normally distributed, the standard scores will also be normally distributed.

Stanines

Stanines are fast becoming the most popular of all standard scores. The word stanine is actually a contraction standing for "standard nine point scale." Unlike the other standard scores, the conversion of raw scores to stanines does affect the shape of the distribution of scores. Stanines are assigned in such a way that the converted scores will always assume the shape of normal distribution.

When raw scores are converted to stanines, the raw scores are changed into one of nine values. The mean score of a stanine distribution is always 5, and the standard deviation is 2 ($\bar{x} = 5$, sd = 2). To make the conversion, the raw scores are rank ordered. These ordered scores are then divided into percentages, in which approximately the lowest 4% of the scores will all receive a stanine score of 1, the next lowest 7% receive a stanine score of 2; the next lowest 12% are stanine 3, the next 17% are stanine 4, the middle

20% are stanine 5. Stanines 6 – 9 consist of exactly the same percentages as stanines 1 – 4, but are in reverse order, and form the higher end of the normal curve. This same distribution can be approximated by first determining the median score, and then calculating the same percentages working out from the middle. This may be preferable since it increases the probability of your distribution being symmetrical around the median.

In Table 2, we present the presumed stanines and their associated scores with their theoretical percentages for a group ranging in size from 25 to 35 subjects. These numbers are approximations. For example, if one took 17% of a group of 25 people to find the number of cases that should be in stanine 4, the number would have to be 4.25. We round off to get an even number of people and to maintain symmetry on both sides of a stanine score of five.

Table 2: Approximate Number of Scores in Each Stanine for Groups of Sizes 25 - 35

Approx. % of scores at each stanine Level	Lowest 4% Level	Lower 7% Level	Low 12% Level	Low Avg. 17% Level	Avg. 20% Level	High Avg. 17% Level	High 12% Level	Higher 7% Level	Higher 4% Level
STANINES	1	2	3	4	5	6	7	8	9
Size of Group									
25	1	2	3	4	5	4	3	2	1
26	1	2	3	4	6	4	3	2	1
27	1	2	3	5	5	5	3	2	1
28	1	2	3	5	6	5	3	2	1
29	1	2	4	5	5	5	4	2	1
30	1	2	4	5	6	5	4	2	1
31	1	2	4	5	7	5	4	2	1
32	1	2	4	6	6	6	4	2	1
33	1	2	4	6	7	6	4	2	1
34	1	2	5	6	6	6	5	2	1
35	1	2	5	6	7	6	5	2	1

Stanine scores are also limited in their usefulness. The most appropriate use of stanines is when the trait being measured can be assumed to be normally distributed. If the trait is not normally distributed the stanines will misrepresent the relative position of scores.

Other limitations are that stanines tend to lose information since more than one raw score may have the same stanine. The more reliable the test, the more likely that this loss of information is meaningful. However, this is not important if one is interested in de-emphasizing small differences. Also, we must consider as a limitation the unfamiliarity of the general population with this form of score expression. This lack of understanding on the part of the public restricts the usefulness of stanines.

Percentiles (P) and Percentile Rank (PR)

Percentiles and percentile rank are among the most common methods used to describe a subject's relative position. They are most familiar to parents, students and teachers so they facilitate communication and are less subject to misunderstanding, assuming everyone really understands these scores.

Percentiles range from 0 to 100, and they are points in a distribution of scores below which a specific percentage of scores falls. If a person's score was at the 25%ile, this means the score was at the point in the distribution of scores where 25% of the people had lower scores and 75% of the people had scores that were higher on that particular test. Since percentiles are points, it would be inappropriate to say that someone falls in a particular percentile because it is impossible for anyone to fall in a point, as a point has no dimensions. However, we can talk about someone falling within a percentile rank since percentile ranks are ranges of scores between percentile points.

Percentile ranks are very closely related to, and are defined in terms of, percentiles. The percentile rank of one (PR1) is assigned to all scores falling between the two percentile points of 0 and 1 (P0 and P1). The PR2 is assigned to all scores that fall within the range of P1 and P2; this procedure continues through the percentile rank of 100.

The major advantage for using percentiles and percentile ranks is the general familiarity of the population with them. Unlike standard scores, the interpretation is not affected by the shape of the distribution. They are interpreted exactly the same no matter how the scores are distributed.

The major disadvantage, and a very important one, is that the difference between percentiles and percentile ranks does not reflect the equivalent difference between raw scores. Percentile ranks tend to hide large raw score differences if they occur at either extreme of the distribution. On the other hand, they tend to exaggerate small differences in raw scores if those differences occur in the middle of the distribution. This distortion is most likely to occur if the scores are normally distributed.

Grade Equivalence

One of the most frequently misinterpreted methods of presenting information concerning a student's relative standing in a group of students is grade equivalence. Grade equivalence is determined by calculating the mean score or median score on a particular test for students beginning each grade level. The average performance on that test at each grade level becomes the raw score equivalent for that grade level. For example, if the average raw score for a large number of beginning third graders is 75, then a raw score of 75 would earn a grade equivalent of 3. If a second or sixth grade student obtained a raw score of 75 on that test, then the grade equivalent for those particular students would be third (G.E. = 3.0).

Grade equivalence is generally reported in two numbers. The first number represents the grade placement and the second indicates the month. For example, if a person had a grade equivalent score of 3.2, this would mean that the same raw score as an average third grade student in the second month of school was obtained.

The most popular use of grade equivalence is in elementary schools. Teachers tend to find them easy to understand and an efficient method for communicating students' performances to parents. In turn, parents tend to find it relatively easy to compare their child's actual grade level with his reported grade equivalent in a particular subject area. This helps them understand their child's performance in comparison to what is expected of him/her at a particular grade level. For most parents, grade equivalence is generally found to be easier to understand and a better communicator of pupil progress than are percentiles or percentile ranks.

Grade equivalence is also used to help teachers detect and diagnose learning problems or subject matter areas of weakness. Teachers can easily identify those students who are performing below what is expected at the particular grade level. They can then provide these students with appropriate help to strengthen the indicated areas of weakness.

Figure 8: Diagram of assumed grade equivalence

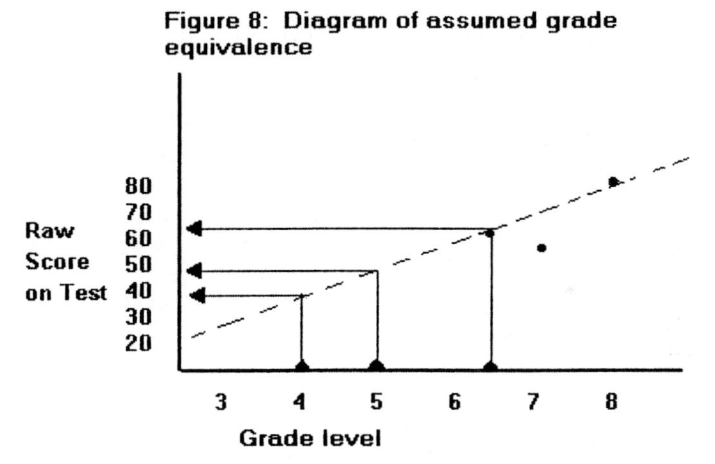

The major limitations of grade equivalence are a function of how they are calculated. The general procedure for calculation is to present a large number of students in consecutive grade levels with a test at the same time. The average student performance at each grade level is then calculated, these averages are plotted, and a smooth line that most closely represents them is drawn (see Figure 8).

In reality, it is not the average point that becomes the grade level but the point where the line passes most closely to the calculated average. For example, the calculated average score for fourth grade may be a score of 50, as it is in Figure 8, but the score of 40 is the reported grade equivalent raw score since the grade equivalent is read off the line and not off the point. It also should be noted that even though no one was tested during the fourth month of the sixth grade, this value can be obtained by locating the point on the X-axis representing the fourth month of the sixth grade and then reading up to where it intersects the line and across to that raw score value on the Y-axis. In this case, a G.E. of 6.4 would have a raw score of 62.

Problems tend to be created because the test is seldom given to all grade levels. However, the line representing grade equivalence is usually extended (as indicated by the dotted line in the preceding figure) to include grade levels not tested. This extension is based on the assumption that learning is a continual process which occurs with regular rates of growth. However, there is empirical evidence that this is not the case, even though it is assumed to be true for children in the elementary grades.

A final word of caution in using grade equivalent scores; while they appear to be so easy to understand, they are frequently misunderstood. A parent who is told that her third grade son's math score is equivalent to an eighth grade level on a particular test frequently interprets this to mean that her son is capable of doing eighth grade work. The correct interpretation is that her son did as well as a beginning eighth grader on that particular test.

It is highly possible that the test did not sample eighth grade mathematics but tested addition, subtraction, multiplication and division. Therefore, the correct interpretation would be that on a test of addition, subtraction, multiplication and division, her son did as well as an average eighth grader would do in this content area. It does not say anything about how well her son would do on an eighth grade test which included fractions, algebra, graphing quadratics, geometry, and other mathematical concepts.

Age Equivalence

Age equivalence is conceptually the same as grade equivalence. It is calculated in the same manner but reported out differently. Where grade equivalence is divided into ten intervals per year, age equivalence is divided into twelve units, for the twelve months of the year. It is reported out by age and month, with a hyphen between the age and month. A student with an age equivalent reading score of 9-11 is reading at the level of an average child who is nine years, eleven months old. Age equivalence has similar advantages and limitations as grade equivalence.

Chapter Summary

In this chapter, we have presented the basic statistical concepts that are most commonly used. Understanding of these concepts is necessary for the intelligent use of tests and grading in the educational setting.

A differentiation was made between descriptive and inferential statistics. It is descriptive statistics that classroom teachers are generally required to deal with including measures of central tendency, measures of variability, standard scores, percentiles and grade equivalence.

The three measures of central tendency are the mean score, median score and mode score. The mean score is the most appropriate measure when the data at least approximates a normal distribution. For skewed distributions, the most representative measure of central tendency is the median score. The mode score is at best a quick, but highly inaccurate, and unstable estimate. In a perfectly normal distribution, these three measures all fall exactly at the same point.

The degree to which scores vary around the measure of central tendency is the variability. Three measures of variability are range, quartile deviation and standard deviation. Standard deviation is most often used and is appropriate, whenever the mean score is used. It, like the mean score, forms the basis of many inferential statistical procedures.

Teachers frequently compare scores from different tests inappropriately. They generally do not consider differences in mean scores and standard deviations on different tests. The best way to avoid this error is to convert raw scores to standard scores.

Standard scores have a common mean and a common standard deviation. One type of standard score is a Z score that has a mean score of 0 and a standard deviation of 1. T scores have mean of 50 and standard deviation of 10. Standard scores allow for more accurate comparisons without changing the shape of the distribution.

Another type of standard score is the stanine, having a mean of 5 and a standard deviation of 2. Unlike the other standard scores mentioned, when changing raw scores to stanines, the shape of the distribution is changed to approximate a normal curve.

Percentiles and percentile ranks are very frequently used to report students' relative performances. Their frequent use is partially due to the apparent ease with which they can be understood. They describe a student's relative position compared to other students. However, a problem exists if the data is not evenly distributed. They tend to exaggerate differences in the middle of the distribution, and they minimize differences at the extremes.

One of most misunderstood methods of reporting a child's performance is grade equivalence. If used, the most appropriate use of grade equivalence is in the elementary school. They are less appropriate in the high school where there is less continuity of subject matter.

The major confusion in reporting grade equivalence is thinking that a fifth grader who has grade equivalence of 8.0 is capable of doing eighth grade work. In actuality it means is that the child is performing that particular task as well as average beginning eighth grader would perform on the same task.

Objectives for Chapter II

After completing Chapter II, you should be able to:

1. Define correlation.
2. Define and interpret the correlation coefficient (r).
3. Define the characteristics of r.
4. Plot a scattergram for a data set.
5. Estimate r for a set of given scattergrams.
6. Define r^2.
7. Define rho.
8. Tell when it would be appropriate to use r, rho or eta.
9. Define reliability.
10. Define three methods for estimating reliability.
11. List ways that reliability is affected.
12. Define validity.
13. List and define five types of validity and state the strengths and weaknesses of each type.
14. To facilitate the achievement of these objectives, see the hyperlinks that are embodied in the text.

Chapter II

Correlation, Reliability and Validity Correlation

Correlation is a measure of the degree of relationship between two or more variables. Variables in this case are defined as: anything one is interested in measuring, including such things as test grades, grade point averages, gender, height, etc. If we are interested in determining the relationship between gender and achievement scores, gender would be one variable and achievement scores the other. The correlation between these variables would indicate if a relationship exists and the magnitude of that relationship. With this information, we could, perhaps determine if male or females have higher achievement scores, on the average.

Correlation Coefficient (r)

The correlation coefficient is the most commonly used measure of correlation. Also referred to as the Pearson Product Moment Correlation Coefficient, or the Pearson r, it is an index that measures the degree of linear (straight line) relationship between two variables. The size of r can range from +1 through 0 to −1.

A correlation of r = 0 means that there is no relationship between the two variables we are interested in measuring. For example, it is likely that r would be zero between the variables of hair color and shoe size. Another way of saying this is that knowing a person's hair color does not allow for guessing that person's shoe size any better than by chance alone. Therefore r = 0.

If the correlation between two variables is +1 or −1, a perfect relationship exists between these two variables. In reality there are very few, if any, perfect relationships.

A positive sign or no sign in the correlation indicates that the two variables are related in such a way that if one increases the other also increases and if one decreases the other decreases. A negative sign indicates a negative relationship that means that as one variable increases the other decreases.

One can predict equally well if the correlation is a positive or a negative. A +.8 correlation is also as predictive as a correlation of -.8. The positive or negative sign only indicates the direction of the relationship and not the magnitude. The magnitude of the relationship is indicated by the number, regardless of the direction. The closer the number is to +/- 1, the greater the magnitude. Thus, a correlation between two variables of r = -.9 would indicate a greater ability to predict than would r = +.8.

An example of positive relationship would be the relationship between the amount of gasoline consumed and the number of miles driven. Another example of a positive relationship would be between a student's achievement score and grade point average. This not a perfect correlation, but it is usually positive. Generally, the higher one's achievement, the higher the grade point average (GPA) (see Figure 9).

Figure 9: Hypothetical Correlation Between Grade Point Average and Achievement

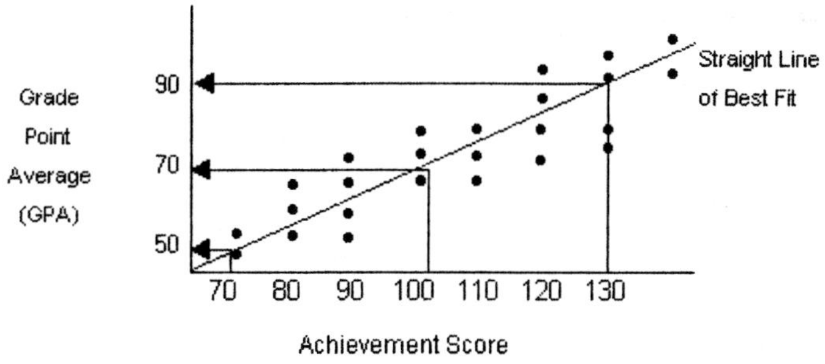

As can be seen in Figure 9, generally, as the achievement increases so does the grade point average. The straight line of best fit in the figure can be thought of as a prediction line or regression line. For example, to make the best guess of what grade point average for someone with an achievement score of 130, one would read up to that prediction line and then across to the grade point average on the vertical axis. For example, when using Figure 9, the best prediction of the GPA for a person with an achievement score of 70 would be a GPA of 50. With an achievement score of 100 the prediction for the GPA would be 70; and for an achievement score of 130, 90 is the predicted GPA. Figure 9 is indicative of a positive relationship because as the achievement score variable increases or decreases so does the predicted grade point average.

An example of a negative relationship is illustrated in Figure 10. This figure shows the relationship between the amount of time it takes to run a mile and age when the only ages being considered are between 3 and 12 years.

Figure 10: Example of a Negative Relationship

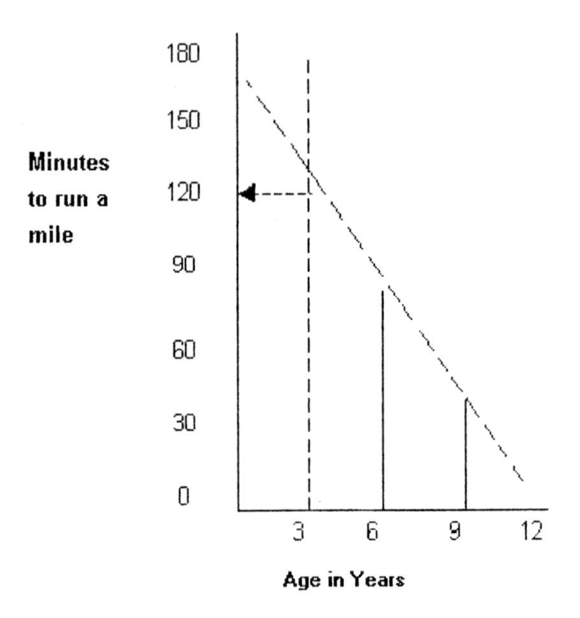

This figure illustrates that as age increases the amount of time it takes to run a mile decreases. The following web page provides additional general information and examples of correlations: http://psychology.wadsworth.com/workshops/correlation.html.

Scattergram

If one is interested in the relationship between achievement and grade point average (GPA) for students, the best way to determine this information would be to plot each student's achievement score against his or her GPA. Let us assume we have twenty students. We would first make a table showing each student's achievement and GPA (see Table 3).

Table 3: Achievement Scores & GPA

S	Achievement Score	GPA
1	130	92
2	70	55
3	80	60
4	80	50
5	130	80
6	120	90
7	120	80
8	110	80
9	110	77
10	120	82
11	110	75
12	100	80
13	100	78
14	100	72
15	100	70
16	100	60
17	100	60
18	90	60
19	90	55
20	100	70

The scores would then be plotted for each student (see Fig. 11).

In Figure 11, three of the plotted points have been numbered for explanatory purposes. Student 1 had an achievement score of 130 and a GPA of 92. Student 2 had an achievement score of 70 and a GPA of 55. Student 12 had an achievement score of 100 and a GPA of 80. All other students' achievement and GPA scores were plotted in the same manner. The result is a figure known as a scattergram.

Figure 11 indicates a positive relationship because the plotted points go from low in the left hand corner to a high in the right hand corner. If the plot went from a high in the left hand corner to a low in the right hand corner, this would indicate a negative relationship (see Figures 12 and 13).

Figure 11: Example of a Scattergram

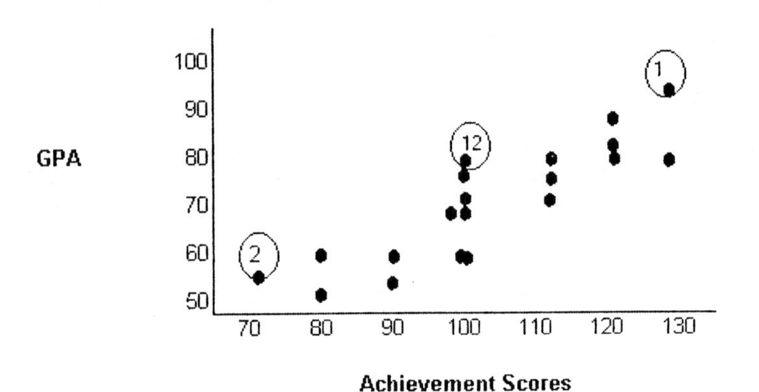

Figure 12: Postive Relationship

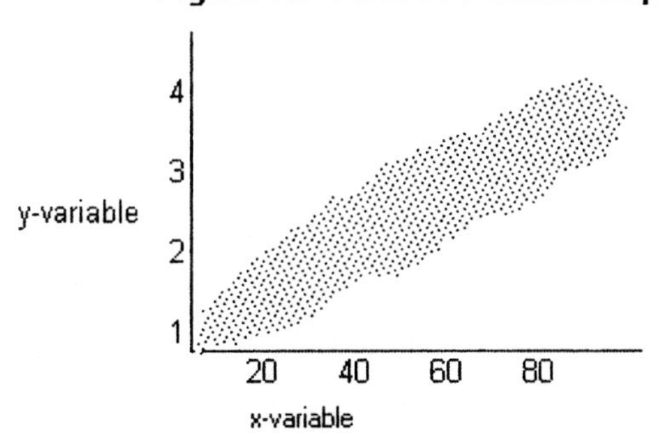

Figure 13: Negative Relationship

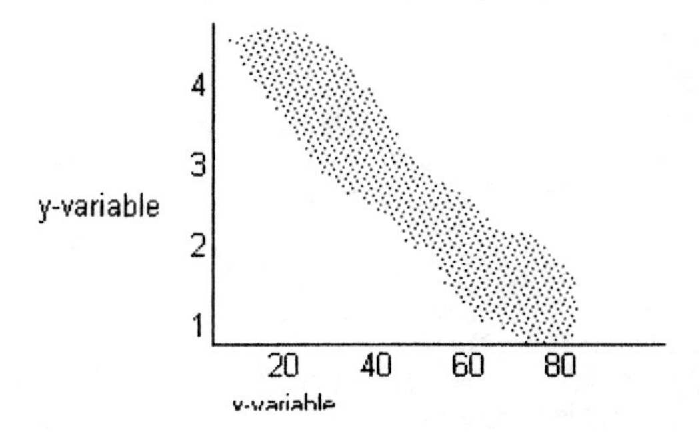

When the correlation is perfect, all plotted points will form a straight line (-1 or +1; see Figure 14). For these perfect correlations one can predict exactly, without any error, by reading from the X-axis to the line and then across to the Y-axis.

Figure 14: Correlation Between Test A and Test B

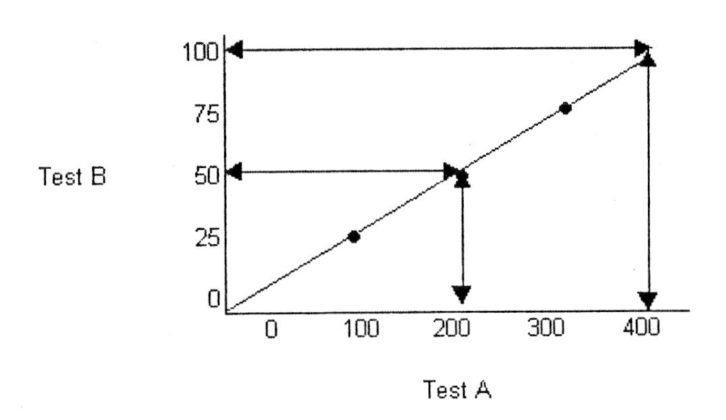

For example, we can predict perfectly that if a person got 200 on test A he would get 50 on test B. We can predict just as well going from B to A as we can from A to B. So if we know a person received a score of 100 on test B, we can predict a score of 400 on test A.

A perfect correlation (r) means that every single point falls on a straight line; no point deviates from that line. As we stated earlier, very rarely, if ever, does this occur.

Figure 15 illustrates a more common situation in which all points do not fall on the line. It is basically a scattergram with a straight line drawn to represent the average of the points. The straight line is calculated in such a way that each point will deviate from it to a minimum degree. The formula for calculating this line is the formula for the correlation coefficient.

The straight line is conceptually the same idea as the mean score. The degree that each point deviates from the line can be conceptually thought of as the same idea as the standard deviation. When dealing with correlation the straight line of best fit is called the regression line and not the mean score, and the degree that each point deviates from that line is called the standard error of estimate instead of the standard deviation.

Figure 15: Example of Illustratin of Points Deviating from the Line of Best Fit

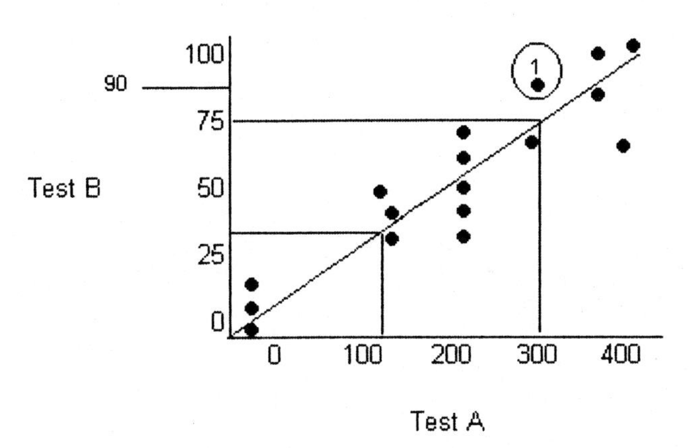

Referring back to Figure 15, if we wanted to predict person I's score on test B from his known score on test A we would read up the regression line from 300 and across to the Y axis. We would then predict his score to be 75 while his actual score is 90. The difference between the predicted score and his actual score is called **error**. If the standard error of estimate is 0, it would mean that no points deviated from the straight line and there would be no error. The more the points deviate from the straight line, the greater the error and the lower the correlation. When the correlation is 0, the standard error of estimate will be at its maximum value. When the correlation is either +1 or −1, the standard error of estimate is at its minimum value, which is zero.

Figures 16, a, b, c, and d present an illustration of four scattergrams with their approximate correlations. As one can see, the closer the points are to a straight line, the higher is the correlation. As the points approach a perfect circle, the correlation approaches 0. These figures only illustrate positive correlations, but the exact same thing would be true for negative correlations.

Figure 16d indicates that no matter how you try to place a straight line to minimize the deviation of scores from the line, no line is better than any other line. When this occurs, we assume that a horizontal line is the line of best fit. The following web page provides an opportunity to simulate the relationship between the shape of a scattergram and a correlation:
http://www.ruf.rice.edu/~lane/stat_sim/ comp r/index.html.
By changing the shape of the scattergram, one can see that the magnitude of the correlation increases as the points approach the line (the standard error of estimate becomes smaller), and that the magnitude of the correlation approaches zero when the standard error of estimate becomes larger (the shape of the scattergram becomes circular).

Interpretation of r

Assume we have a situation as presented in Figure 17, and we know that the standard error of estimate (Sxy) is equal to 10. We can then say that in 68% of the cases the actual score the individual received will be within the limits of the predicted score ± 1Sxy. In the example on the next page, the predicted score for subject 1 is 55. The Sxy is 10. Therefore, we can assume that 68% of the time the actual score (in this case 58) will fall between 55 – 10 and 55 + 10 (45 to 65). We can also assume that 95% of the time the actual score will fall within the limits of the predicted score plus and minus 2Sxy. In the case of subject 1 we can assume 95% of the time his or her actual score will fall between the predicted score of 55 and ± 2Sxy, or between 35 and 75. This concept is the same as the concept discussed in Chapter 1 for the mean score and standard deviation.

Figure 16. Examples of Four Scattergrams and Their Approximate Correlations

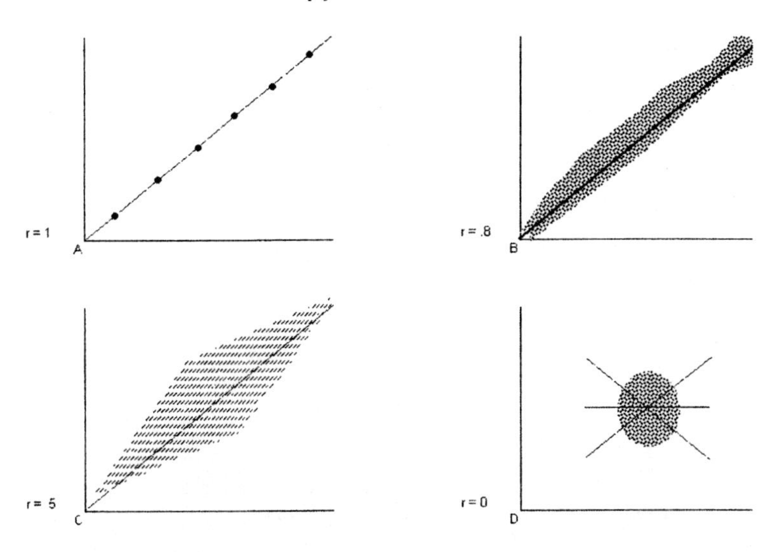

Figure 17: Scattergram with Regression Line

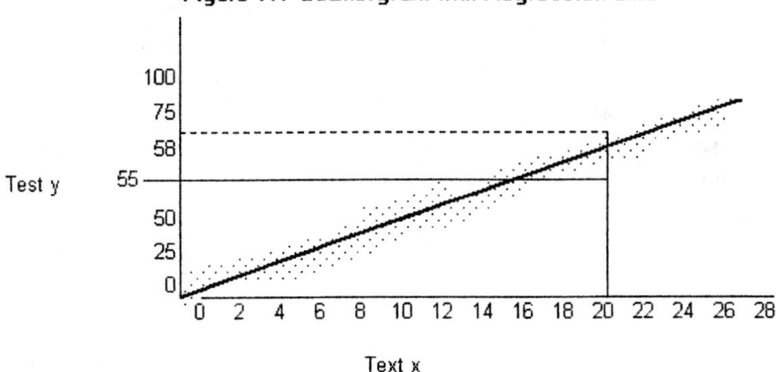

The smaller the standard deviation and the standard error of estimate, the more accurate are our predictions. In the case of r, the closer r is to 1, regardless of its sign, the smaller is the Sxy. Therefore, our prediction increases in accuracy as the correlation increases. (See Chapter 3, Table 7 for examples of how to use a table to determine if r is significant.) The following web page provides an animation that allows one to see the effect of changes in the standard error of estimate on the correlation between two variables:

http://www.ruf.rice.edu/ ~lane/stat sim/comp r/index.html.

This animation will also allow changes to the standard deviation of the independent variable. By changing the standard deviation and the standard error, one can change the shape of the scatter plot and, therefore, the correlation between x and y.

r^2

The best way of interpreting r is in terms of the percentage of variance it accounts for. This can best be explained by an example. The correlation between the College Board of Examination and success in college is r = .4. If we square r, that is .4 x .4 = .16 (or 16%), this will tell us how much variance is being accounted for when we use the College Boards to predict success in college. In this case only 16% of the reasons someone does well or does poorly on the College Boards are the same as the reasons someone does well or does poorly in college. Therefore, 84% of the reasons (variability) (100% - 16% = 84%) for success in college are not being accounted for by the College Boards. Even if the correlation is highly significant, that is, it can predict better than chance (better than 50-50 chance), to interpret it we must know more than its level of significance. One must also know the amount of variance accounted for, which is calculated by squaring r.

Rho

Rho is really a special case of r, and it is generally an underestimate of r. Theoretically, it should be used in the data is interval (that is, it is obtained from a measuring scale where there are assumed to be equal intervals between the points on the scale), but can assume it is ordinal (that is, it is obtained from a measuring scale where the points on a scale are assumed to be rank ordered, but there are not necessarily equal intervals between the points). Rho is generally easier to calculate than r when the number of subject pairs is relatively small. If the N is larger than 20 or 25, it then becomes difficult to calculate by hand.

A caution when interpreting r or rho is that one must keep in mind that r and rho are measures of straight-line relationships. If the relationship is curved, then r and rho would be an underestimate of the true relationship. For example, there is experimental evidence that a curved relationship exists between anxiety and performance. That is, people with low and high anxiety levels have poor performance while people with anxiety have higher performance (see Figure 18).

If one calculated an r or rho, the r and rho would be 0 indicating that there is no linear relationship. This does not mean that there is no relationship nor does it mean that we cannot predict perfectly. In the cases presented in Figure 18, there is perfect prediction. Low and high anxiety scores for this individual produce low performance, and medium anxiety produces high performance. In this case and others where one expects a curved relationship, the correct correlational technique to estimate the relationship would be the correlation ratio, also called eta (η).(1)

Eta measures the degree of relationship but does not give direction. When the best fitting line is a straight line, $R^2 = \eta^2$. If the best fitting line is not straight, then R will underestimate the relationship ($R^2 < \eta^2$).

Reliability

Correlation is very frequently used to estimate reliability. The reliability of a test is defined as the consistency of the measure. This means that the test, no matter what it is measuring, will produce the same value or one very close to it every time it is used.

Figure 18: Curved Relationship

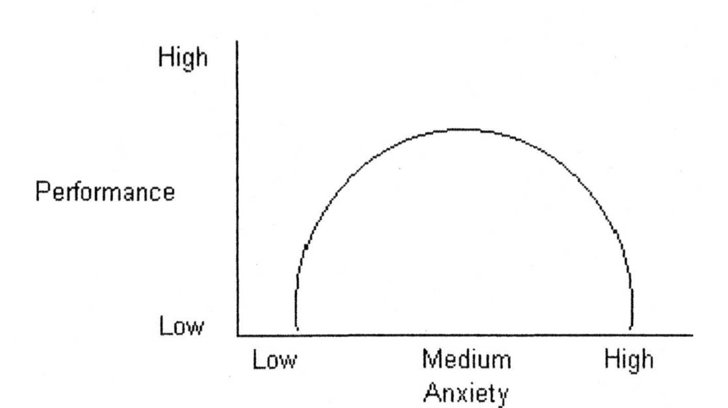

Methods for Estimating Reliability

The consistency of a test is generally expressed in terms of r. If r is equal to 1, the test is considered to be perfectly reliable. If r = 0, the test is totally unreliable. If r is negative, the reliability of the test is difficult to determine because negative relationships should not occur when dealing with reliability. In general, the reliability coefficient should be .9 or higher before a test is considered reliable.

Test-Retest Reliability

One way of determining reliability is through the test-retest method. This necessitates giving the subjects the test and then retesting the same subjects with the same instrument at a later date. If they get the same or nearly the same score both testing times, then the test is considered to be reliable. However, if the scores differ widely the test is considered unreliable. The major shortcoming to this method of reliability measurement is that since the subjects are being retested on the same items, they may remember some of the items from the first testing. This recall may help them improve their performance during the retest that will decrease the test reliability. Another shortcoming is that the longer the time period between testing, the greater the likelihood that differential or new learning will occur. This learning would also affect test performance and therefore reduce reliability.

Equivalent Form Reliability

The equivalent forms method reduces the problem of recalling previous questions. This method is based on the assumption that one has two equivalent forms of a test that measure exactly the same thing. One popular method of constructing these equivalent forms is to construct twice as many items as are needed for one test and then randomly assign them to either form A or form B.

Each subject receives both test A and B. One-half of the subjects are tested on test A first then test B, and the other half are given test B first and then test A. The score on both forms for each subject are then correlated and produce an estimate of the reliability for the test.

Internal Consistency (Split-half and Kuder-Richardson)

Unlike the other two methods of estimating reliability the internal consistency method does not require two testings. It also depends on correlation to estimate its reliability, but it correlates the test items within the same test. That is, if one gets a high estimate of reliability using internal consistency, this means that all of the items on the test are basically measuring the same underlying concept.

The two major methods of calculating reliability using internal consistency are the split-half method and the Kuder-Richardson formula 20 (KR 20). When using the split-half method, the items of the test are dived in half. Frequently, this is accomplished by splitting the odd and even numbered items, or almost any other split can be used, except splitting by grouping the items in the first half of the test against those in the second half of the test. The number of correct items are then calculated for each of the two groups of items, and these are correlated to arrive at a value of reliability for that test. The major problem with this technique is that it estimates reliability based on half of the number of items. Since reliability increases as the number of items increase, this method underestimates the reliability of the test. To correct this problem, the Spearman-Brown Correction Formula is usually applied. Generally, you do not need to know how to do this calculation, but you should be familiar with the term if you read it.

KR 20 also provides an estimate of reliability based on a single administration of a test. Conceptually it averages the correlation between every item and the overall test. The higher this average is, the more reliable the test. If the test is measuring more than one underlying concept, so that every few items are measuring unrelated concepts, then the KR 20 will tend to produce a low reliability coefficient even though the same test may produce a high reliability coefficient using test-retest, equivalent forms, or even possibly split-half. Even with this major limitation, primarily because of its convenience and statistical soundness, KR 20 is becoming the most widely used estimate of test reliability. It is being used to report the estimates of reliability of standardized tests.

Aspects that Affect Reliability

The reliability of a test can be altered by any one of the following factors:

1. Increasing the number of items on a test will increase the reliability.
2. Objective methods of scoring will increase test reliability.
3. Having a test measure one particular concept is likely to increase reliability. However, the items should not be interdependent.
4. Constructing the items so that they are approximately equivalent in item difficulty will increase reliability.

5. Tests administered in a standardized manner will have higher reliability.

One must remember that saying a test is reliable only means that it will consistently produce the same rest results for a respondent. It does not mean that if we constructed an intelligence test with a high reliability coefficient that we would actually be measuring intelligence. For example, we could construct a test to measure intelligence and use height as the indicator of intelligence; that is, the taller the person the more intelligent. Our test would probably be highly reliable, especially for adults, since their height is not likely to fluctuate between testings. Therefore, we could have a test that is highly reliable because we can consistently get the same values. In essence, reliability alone will not guarantee a good test, but without it the ability to interpret and use a test will be greatly limited because the inconsistent results will highly increase the error in prediction.

Validity

The most important characteristic of any test is its validity. The validity of a test is commonly defined as the degree to which a test measures what we want it to measure. For example, if we constructed a test to measure subjects' mathematical abilities, the validity of that test would be the extent that the test was capable of measuring the students' actual mathematical abilities. One never really knows the true validity of a test; we only estimate. There are many methods for estimating validity. Often one method has more than one name. The five types of validity that will be discussed here are face, content, concurrent, predictive, and construct validity.

Face Validity

The least accurate estimate of a test's validity face validity. It is arrived at gauging by the subjects' reaction to the test. If the test appears (to the subjects) to be measuring that which it purports to measure, it has face validity. One type of face validity may be *expert judge validity* that is a little more sophisticated. In this case, content experts instead of subjects judge the validity. For this type of validity, an actual correlation coefficient is not calculated.

Content Validity

Content validity has also been called definition validity and logical validity. Generally, the content validity of a test is estimated by demonstrating how representative the test items are of the content or subject matter the test purports to measure. The correlation coefficient is calculated by considering the subject mater as one variable and the test items as the second variable.

Concurrent validity

Concurrent validity is estimated by how well a test correlates with another test that has already had its validity estimated. For example, if we are interested in constructing an intelligence test we can estimate its validity by correlating it with well-established intelligence tests such as the Stanford-Binet or the Wechsler Intelligence Scale for Children (WISC). If our test correlates highly, generally r greater than .7, with the other tests we may reasonably assume that it has concurrent validity.

Another way of estimating the concurrent validity of a test is by identifying people who have been assumed to be highly intelligent and people who have been assumed to be low in intelligence. We could then administer this test and see if the people who have been identified as highly intelligent score high, and those identified as low in intelligence score low. This kind of validity is called *known group validity,* and is generally considered to be a type of concurrent validity.

Predictive validity

The major difference between concurrent and predictive validity is that in predictive validity one predicts into the future on the basis of the test results and then checks that prediction. If the prediction is correct, or if one can predict significantly better than by chance, the test has predictive validity. The higher the correlation between the test and the outcome, the better the predictive validity. Concurrent validity, on the other hand, may look at the same types of relationships but it is not predicting into the future; it is more present oriented. That is, the correlation between the test and its criterion occur at approximately the same time. Predictive validity and concurrent validity, when taken together, have been called *criterion validity, empirical validity,* and *statistical validity*. All three terms are basically describing the same characteristic.

Construct validity.

Construct validity is a conglomeration of all other types of validity. It is most important when one is interested in interpreting a test score as a measure of some particular construct or attribute, such as intelligence, creativity, neurosis, etc.

When one is interested in construct validity, there is generally no one criterion that is acceptable as a measure of the construct. There is most likely a series of criteria that the test should relate to in differing degrees. How this test should relate to the various criteria is dictated by the theory of the construct. For example, someone may have a theory of intelligence that may include academic success, financial success, relatively good scores on the other intelligence tests, and popularity among co-workers. The theory may further predict that the relationship between the test and the above mentioned criteria may be the following:

1. The test should correlate highest with economic success and popularity.

2. There should be moderate correlation between the test and academic success.
3. The lowest correlation will be between the test and other intelligence tests.

The test could then be correlated with the above criteria. Construct validity exists to the degree that the results are in the predicted directions.

Usefulness of Validity

Obviously, the above methods of estimating validity vary in their degree of usefulness. Any estimate of validity is generally better than no estimate, but there are times when no estimate is better than poor estimates because you think you have validity when you do not. This may lead to a false feeling of security in your test, and you may make totally unjustifiable decisions based on your test.

When estimating the usefulness of a particular test, it is important to consider the type of estimate of validity reported. Many of the most widely used standardized reading tests generally report only content validity. Even tests that report under the heading of construct validity are frequently inappropriately titled. A close examination generally reveals that they have used no more than content or face validity. Since content and face validity are unrelated to the test's ability to predict, it would be inappropriate to consider a score achieved on these tests as an indicator of probable success. However, these tests are frequently used as indicators of success. This does not mean that the tests are unable to predict success, but since the validation procedures that were used did not estimate its predictive ability, one has no estimate of how accurately the tests can indeed predict.

Probably the most useful types of validity are predictive and concurrent. However, construct validity is the best of all but it is most difficult to ascertain. If one has construct ability, one also has predictive, concurrent, and content validity.

It is important to keep in mind that we never really have a totally valid test. We can only estimate its validity; and because of this, there are always errors in prediction. If one can be sure that a test has a high estimate of validity, reliability estimates will be less important. One can have reliability without having validity; but if the test is valid, it is also reliable. It therefore follows that most of the time spent in test construction should be devoted to improving validity and not the reliability. This is contrary to most common practices. The following link will provide additional examples of the concepts of reliability and validity:

http://www.wadsworth.com/psychology_d/templates/student_resources/workshops/reliability1.html.

Chapter Summary

To determine the degree of relationship between variables we use a technique called correlation. The most common correlational technique for linear correlations is the Pearson Product Moment Correlation (r). It ranges in magnitude from +1 to −1 with r = 0 indicating no relationship at all.

An easy method for determining the degree relationship between variables is plotting a scattergram. The closer the points approximate a straight line, the greater the degree of correlation.

To interpret a correlation at least two things have to be considered: the standard error of estimate (Sxy) and the proportion of variance accounted for (r^2). The standard error of estimate is conceptually similar to and is interpreted in the same manner as a standard deviation (sd). r^2 provides information on the proportion of the variance that can be accounted for by the correlation.

A major caution when using and interpreting r is that it is a linear technique. If the relationship is not linear, other techniques such as eta (η) should be used. Rho is appropriate to use when the N is small, the relationship is linear, and you cannot make the assumption that the data are at least interval. In other words, the data are ordinal (rank ordered).

Correlation is very frequently used to measure a test's reliability. There are three methods for estimating reliability. These are test-retest, equivalent forms, and internal consistency. Each of these methods has its strengths and weaknesses. One method of measuring internal consistency that has become very popular is the Kuder-Richardson (KR 20).

It is important to keep in mind that reliability is necessary but not in itself sufficient for the meaningful interpretation of a test; we must also know the test's validity. Validity is the most important aspect of any test. It is the ability of a test to measure the trait or attribute for which it was designed. The procedures to estimate validity can generally be classified into five areas: face, content, concurrent, predictive, and construct validity.

Concurrent and predictive validity together have been called criterion validity. Generally, this is the most useful and practical estimate of validity. Construct validity is made up of all of the other types of validity. It is the best estimate of validity, but is most difficult to obtain.

In construction or choosing of a test, one should determine the type of validity used. One then must determine if this is appropriate for the intended use of the test.

---

(1) more detailed explanation of eta can be found in Nunnally, J.C. Psychometric Theory (2nd ed.), New York: McGraw-Hill Book Co., 1978, pp. 146-150; and Lomas, R.G., Statistical Concepts. New York: Longman, 1992, pp. 31-32.

Objectives for Chapter III

After completing Chapter III, you should be able to:

1. Define statistical significance.
2. Distinguish among the .05, .01, and .001 levels of significance.
3. Define what a null hypothesis is.
4. Define and state the relationship between a Type I error and the alpha (α) level.
5. Define and state the relationship between a Type I error and Type II error.
6. Define statistical power.
7. Define a one-tailed test and a two-tailed test and state when it is appropriate to use each.
8. Identify tests of statistical significance and tell when it is appropriate to use them.
9. Define robust.
10. Explain the meaning of the statistical statement, $p \leq \alpha$.
11. Look up a given *t*, $\chi 2$, F and r in a table and determine if they are statistically significant at the .05, .01, and .001 levels.
12. Determine how large the *t* or r value has to be for a one, or two-tailed test, for $\alpha = .05$ and $\alpha = .01$.
13. State the similarities and differences between *t* and F –tests.
14. See hyperlinks.

Chapter III

The Meaning of Statistical Significance

Just because a research study finds significant results does not mean that a difference exists. This is a wrongful inference. In reality, we cannot know if the difference is large, that is, the magnitude of the effect, just by the statistical significance. One would also have to know the probability level (alpha, α), the size of the sample (N), the mean (M) and standard deviation (sd). Only then does one have the information to intelligently interpret the usefulness and meaning of the significant finding.

When something is statistically significant, it is unlikely to occur by chance. **Chance** is operationally defined by some alpha (α) level. In psychology and education, the alpha levels that are generally used are .05, .01, and .001. The alpha level you pick is a *subjective* decision. It may be one of these three or any other. However, if you choose a non-standard alpha level, you should be prepared to justify your choice.

Many researchers choose to set the alpha level at .05. By doing this, they are conceptually saying that they are willing to have the relationship between the variables that they are testing occur by chance five times out of 100. With the alpha of .05, if the relationship is statistically significant, the researchers conclude that the relationship that they found is unlikely to have occurred by chance alone. They are conceptually willing to assume that the relationship really does exist, and is not just a chance occurrence. However, it is important to remember that from this information they cannot tell the magnitude of the relationship, only that there is a relationship.

If instead an alpha level of .01 (α = .01) is decided upon the data are found to be statistically significant, then the relationship of these variables is only likely to occur by chance 1 time in 100. If a relationship is found to be statistically significant at α = .001, then it will only occur 1 time in 1000 by chance alone.

If one converted these decimal values to fractions, it is more easily seen that (.05 = 5/100, .01 = 1/100, and .001 = 1/1000). If something is found to be statistically significant at the alpha level .05, the conclusion that a difference exists in the sample of the population one hopes to infer to is likely to be wrong 5 times out of 100 (5/100). At the .01 level one would be 1 time in 100 (1/100), and at the .001 level one would be in error 1 time in 1000 (1/1000).

One should also be aware that by increasing the number of subjects one increases the probability of finding statistical significance, even though this does not necessarily change the magnitude of the relationship. For example, if five boys took an achievement test and had a mean score of 105 with a standard deviation of 15, and five girls had a mean score of 100 with a standard deviation of 15, these two groups would not be found to be statistically different at an alpha level of .05 or at any of the more stringent levels (.01, .001, etc.). However, given the same mean scores and standard deviations (magnitude of effect), but increasing the number of subjects from 5 to 1000 for each group, would yield statistically significant results. The probability of this occurring by chance alone would be less than .05, and would therefore, be statistically significant even though the magnitude of effect had not changed (in both case, boys were 5 points higher than girls, on the average.). Therefore, increasing the N size inevitably increases the probability of finding statistical significance.

From what has been said, it is obvious that someone says that something was found to be statistically significant, that information is not very meaningful unless we know the alpha level, the N size, and the effect size.

Hypothesis

A hypothesis is a prediction statement that a researcher makes. This statement consists of the variables, and their predicted relationship. It is what the researcher then statistically tests.

A null hypothesis is a statement of no difference, no relationship between the variables of interest. Conceptually, the researcher is testing the null hypothesis unless he/she has specifically stated that he/she is testing a directional hypothesis. Most researchers usually want to find that there is a relationship between their variables of interest. To appropriately use statistics, they would predict that no relationship exists, thereby writing a null hypothesis. They then set out to run their studies, and do their statistics, so that they can reject the null hypothesis. If they find statistical significance within their data, they can then reject the null hypothesis. While most researchers develop and test a null hypothesis, many make predictions as to what they will find.

A directional hypothesis is a statement of difference, with a particular relationship between the variables of interest. When a researcher has a directional hypothesis, he/she proposes that there is a specific way that the variables are related or vary.

Type I Error

A Type I error is related to the likelihood (probability) of rejecting the statement that says there is no relationship or no significant difference between groups when that statement really is true. In other words, a Type I error is the probability of rejecting the Null Hypothesis when the Null Hypothesis is true.

The probability of making a Type I error is determined by your alpha level. If you set your alpha level at .05, the probability of making a Type I error is 5 times in 100. Similarly, if an aipha of .01 is chosen, the probability of making a Type I error is 1 in 100.

Type II Error

A Type II error is not rejecting the Null Hypothesis when you should. In other words, there really is a relationship and/or a significant difference between groups, and you fail to reject the Null Hypothesis that says there is no difference.

The likelihood of making this error is inversely proportionate to the likelihood of making a Type I error (as the probability of making a Type I error increases, the probability of making a Type II error decreases). If you hold your sample size constant and make your alpha level more stringent by going from α = .05 to α = .01, you decrease the probability of making a Type I error from 5 times in 100 to 1 time in 100; but you increase the probability of making a Type II error.

A Type II error is related to the power of your test. The *power* of a test is defined as the probability of detecting a difference when one exists. In a more powerful test, you are less likely to make a Type II error.[1]

One-Tailed and Two-Tailed Tests

Prior to computing a test of significance, one must decide whether to use a one-tailed or two-tailed test. A *two-tailed test* is actually a *non-directional* test. This means that the direction of the relationship(s) being tested is (are) not predicted prior to running the analysis.

For example, if we hypothesize that a significant difference exists between achievement scores of males and females, this hypothesis is non-directional because we did not state whose achievement scores would be higher; we just said they would be different.

However, when you have prior data indicating the direction of the relationship, it is more appropriate to use a one-tailed test. A *one-tailed test* is a *directional* test. An example of a one-tailed hypothesis would be...Male achievement scores will be significantly higher than female I.Q scores. This is one-tailed because we are predicting who will be higher. If the prior data or theory indicated that female achievement scores should be significantly higher, and if our hypothesis stated that...Female achievement scores will be significantly higher than male achievement scores; this too would be directional because we are predicting who will do better on the achievement test.

A one-tailed test is more powerful than a two-tailed test. It is more likely to detect a difference or a relationship if one exists in the hypothesized direction. However, if a very strong difference occurs in the opposite direction, you must state your hypotheses have failed to be substantiated and that your results are *not* significant.

The same data *cannot* then be used to compute a two-tailed test of significance. The study should be replicated on a new sample and again you must decide whether to compute a two-tailed test or a one-tailed test in the same direction as before or in the opposite direction. If you are really not sure of the direction of the results, you are obligated to use a two-tailed test. Additional information and examples regarding hypothesis testing can be found at the following of web page:

http://www.wadsworth.com/psychology_d/templates/student_resources/workshops/hypothesis1.html.

Tests of Significance

We have previously discussed levels of significance and we have alluded to tests of significance. In this section, we hope to briefly introduce the most common tests of significance used and show some of the relationships between these tests.

As you may recall from an earlier presentation (pp. 18-21), Z scores were defined as the score, minus the mean score, divided by the standard deviation ($Z = (X-M)/SD$). If we found a person's achievement test score had a Z score of 3 ($Z = 3$), this means the achievement score was three standard deviation units above the mean score of the group. The *probability* (p) that someone would score three standard deviation units above or below the mean score by chance alone would be 1 time in 100 ($p = 1/100 = .01$). For something to be found statistically significant, the probability (p) has to be less than the subjectively decided upon alpha level. In the previous example, if we set our α level at .05, and we wanted to determine if this person's achievement score was significantly different from the average, we can say it was, because the probability of his score being that high or higher by chance alone was less than the stated α level. If the calculated probability (p) is less than or equal to the subjectively decided upon α, ($p \leq \alpha$) then the finding is said to be statistically significant. If the alpha level is set at .01 ($p = .01$), then the results would still be significant since the probability was equal to alpha. However, if we set our alpha level at $\alpha = .001$ and $p = .01$, then this would not be significant since it is likely to occur by chance 1 time in 100 and we will not accept anything as significant that is likely to occur by chance more than 1 time in 1000. In this case, the probability is greater than the alpha level and is, therefore, not significant.

t-Tests

Generally, *t*-tests are run to determine if *two groups* are statistically significantly different. The *t*-test cannot determine if more than two groups are significantly different. It also assumes that the data being analyzed is interval. If one looks at the *t*-test formula, which is generally one sample mean score (M_1) minus another sample mean score (M_2) divided by the standard error of the difference between means, (SD_M, which is conceptually a standard deviation) then $t = (M_1 - M_2)/ SD_M$ is very similar to the Z score, $Z = (X - M_1)/SD$.

There are tables of significance for *t*-tests. After calculating the *t*, one looks up the *t*-value in the table to determine if it is significant. If the probability value associated with the calculated *t*-value is less than or equal to the alpha level ($p \leq \alpha$), then the *t*-test result is considered significant at that α. The following link leads to a web page that will perform a t-test for both independent and dependent (correlated) examples: http://faculty.vassar.edu/lowry/tu.html.

How to Use a *t*-table

The following is an example of how to use a *t*-table. Let's assume two treatments for teaching reading are being studied. There are 16 people in Treatment 1 (n_1=16) and 16 people and in Treatment 2 (n_2 = 16, N = 32). The research hypothesis states the two groups, but it does not predict the direction; so we have a two-tailed test. The alpha level is .05 ($\alpha = .05$) and the degrees of freedom (df) = N-2, which equals 32 – 2 = 30. If the calculated *t* of the *t*-test is as large as or larger than the table value corresponding to the .05 level and 30 degrees at freedom, then the two groups are significantly different at the .05 level. According to Table 4, this value is 2.04. It is the value found at the place where df (30) and $\alpha = .05$ for a two-tailed test intersected.

If we made a directional hypothesis using the same example and stated that Treatment 1 will do better than Treatment 2 (and we find that Treatment 1 did do better), we would check the table for the .05 level for a one-tailed test with the same df = 30. In this case, the t we calculated would have to be as large as or larger than the table value of 1.72 to be significant at the .05 level. If, on the other hand, Treatment 2 was larger, we then have to conclude that there is no significant difference. We must come to this conclusion no matter how great the difference is between the treatments, because the difference is in the opposite direction of that which was predicted.

Table 4: Example of a t-table

df	Level of Significance for a One-Tailed Test				
	.05		.01	.005	
	Level of Significance for a Two-Tailed Test				
	.10	.05	.02	.01	.001
1	6.3	12.7	31.80	63.70	636.7
5	2.0	2.6	3.40	4.10	6.9
10	1.8	2.3	2.80	3.20	4.6
20	1.74	2.1	2.52	2.81	3.9
30	1.72	2.04	2.50	2.80	3.6
60	1.68	2.00	2.42	2.70	3.5
120	1.66	1.99	2.40	2.61	3.4
	1.64	1.97	2.30	2.6	3.3

Note: df = N-2 is used for a test when the two groups being tested are **not** correlated (independent) when N is the total number of subjects in one study. Also note the table values are approximations.

Chi-Square

Chi-Square (χ^2) is a test of significance usually used when one is interested in testing to see if the frequency of occurrence is significantly different between groups. It can be used when dealing with nominal data such as frequency counts.

Chi-Square (χ^2) is calculated by taking the square of the calculated sum of the expected scores (E) minus the observed scores (O), and dividing by the expected score $\{\chi^2 = \Sigma[(E-O)^2/E]\}$. The expected score is conceptually similar to a mean score.

Similar to *t*-tests, you look up the χ^2 in a table to see if the probability level associated with it is less than or equal to the selected alpha level ($p \leq \alpha$). If it is, the χ^2 is significant and one would be able to conclude that the differences in frequency of occurrences between groups is not likely to be due to chance at that specified alpha level. The following web page will calculate the χ^2 (and phi coefficient) for two dichotomous variables: http://faculty.vassar.edu/lowry/tab2x2. html.

How to Use a χ^2 Table

The following example demonstrates how use a χ^2 (Chi-Square) table. Let's suppose someone tossed a coin and came up with heads 40 out of 50 tosses. We want to determine if this is significantly different than what we would expect by chance when a $\alpha = .05$. In this case, there is only one sample, one coin.

There are two cells, one heads and one tails. The value is the observed toss of 40 heads and 10 tails. Below the diagonal is what we would expect by chance –25 heads and 25 tails. In such a case, the degrees of freedom is equal to the number of cells minus one, which in this case is equal to 2-1=1.

Illustration 1: How to use a Chi-Square Table

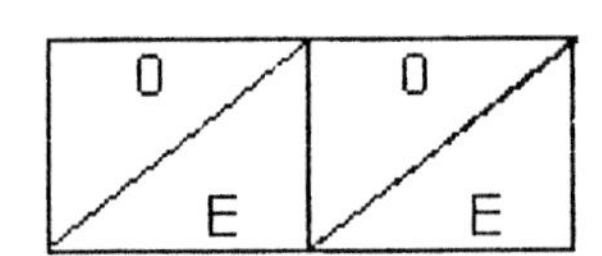

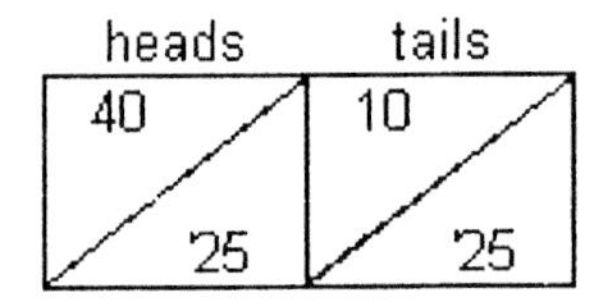

To determine if the calculated χ^2 is significant at the .05 level, one would go to Table 5 and look at the value at the point where .05 and 1 df intersect. If this calculated χ^2 is as large or larger than the table value of 3.9, then the χ^2 is found to be significant. This would mean that it is unlikely that a coin would come up to 40 heads and 10 tails on 50 tosses. These results would occur no more than 5 times in 100 by chance alone, and one could assume the coin was probably biased.

This indicates that if a calculated χ^2 is greater or equal to the χ^2 value listed in the table, then the probability of obtaining a χ^2 this large would be less than or equal to the previously set alpha (not likely due to chance). Therefore, the χ^2 could be statistically significant, one could reject the Null Hypothesis (there is no significant difference), and accept the research hypothesis (there is a statistically significant difference). If the calculated χ^2 is less than the Table χ^2, then the opposite would be true.

χ^2 calculated $\geq \chi^2$ table value: $p \leq \alpha$ statistically significant; reject Ho, accept research hypothesis

χ^2 calculated $< \chi^2$ table value: $p > \alpha$ not statistically significant; do not reject Ho; reject research hypothesis

In another case, let's assume we want to find if there is a significant relationship between political parties and income (the data are nominal). Since, in this case, we have more than one group to compare (political parties and income), the df = the number of columns minus one, times the number of rows minus one [df = (C - 1)(R - 1)]. We have three columns (C=3) and two rows (R=2) so df = (3-1)(2-1) = (2)(1) = 2.

Table 5

Example of a Chi-Square Table

	Level of Significance of a Two-Tailed Test		
df	.05	.01	.001
1	3.9	6.7	10.9
2	6.0	9.2	13.9
4	9.5	13.3	18.5
10	18.4	23.2	29.6
20	31.4	37.6	45.4
30	43.8	50.9	59.7

Note. The df for a χ^2 in which there is only one sample is df = [(#C)-1] where #C is the number of cells. The df for χ^2 in which there are more than one sample to compare: df = (C-1)(R-1) where C is the number of columns and R is the number of rows. Also note that the table values are approximate and tend to be somewhat conservative.

Assume we are interested in determining if there is a significant difference between political party affiliation and income at the .01 level. If the calculated χ^2 is as large as or larger than the table value of 9.2 which is associated with 2 df and the .01 level, then there is a significant difference between political affiliation and income at the .01 level.

Illustration 2: Example of a Chi-Square Data Table

		Political Parties		
		Democrat	Republican	Other
Income	High			
	Low			

F-Tests

The F-test can be used to determine if there is a significant difference between two or more groups or variables, simultaneously. An F-test is the most frequently used test of significance. It is defined as the mean square between groups (MS_b) divided by the mean square within groups (MS_w) and is conceptually similar to differences in means between groups. The mean square within groups (MS_w) is conceptually the error in your data or how much variation you can expect by chance alone and is similar to the standard deviation. Therefore, $F = MS_b/MS_w$.

Like the others, an F is calculated and looked up in a table to determine if the probability associated with that F is less than or equal to the set α. Once again, if it falls within these bounds, then one can conclude that the groups are significantly different at that specified alpha level.

How to Use an F-Table

The following is an example of how to use the F-table. Since the F-test can test for differences between two or more groups simultaneously, the F-test requires two sets of degrees of freedom, one for the numerator and one for the denominator. The degrees of freedom-numerator (df_N) is equal to K-1, where K equals the number of groups. The degrees of freedom-denominator (df_D) is equal to N-K, where N equals the total number of subjects and K is the number of groups.

Let's assume you are interested in determining if there is a significant difference at the .05 level between ten methods for teaching reading (K = 10). There are 11 subjects in each of the ten groups (N=110). The df_N = K-1 =10-1 = 9, and df_D = N-K = 110-10 = 100. The table value associated with these two sets of df (9 and 100) is 1.90 (see Table 6). It is the value found at the place where df_N (9) and df_D (100) intersect. If the calculated F is as large or larger than 1.90, then there is a significant difference between the ten groups at α = .05. This doesn't tell you where the difference between the groups is, just tells you that there is one somewhere. One may want to then run separate *t*-tests between the groups to locate the difference.

If the df listed in the table value does not exactly match what you have for your specific problem, you can interpolate (approximate).

Table 6: Example of an F-Table

df_D	1	5	9	20	30	100	--
1	161	230	242	248	250	253	254
10	5.26	3.40	3.0	2.80	2.70	2.60	2.60
30	4.17	2.60	2.16	1.93	1.84	1.69	1.62
60	4.00	2.40	2.00	1.80	1.70	1.50	1.40
100	3.94	2.30	1.90	1.70	1.60	1.40	1.30
200	3.90	2.30	1.90	1.70	1.60	1.40	1.30
400	3.89	2.30	1.90	1.60	1.50	1.30	1.20
--	3.88	2.29	1.83	1.58	1.47	1.25	1.00

Note: The df for the numberator (df_N) is K-1 where K is the number of groups to be compared. The df for the denominator (df_D) is N-K where N is the total number of subjects in the study and K in the number of groups. Also note the table values are approximate for *a* = .05 only.

Similarities and Differences between the F-test and the t-test

1. Both tests give probability statements about statistical significance.

2. A *t*-test can only test the difference between two groups at one time.

3. Both tests require interval data as an underlying assumption.

4. t^2 is equal to F when you are testing the differences between two groups.

5. F-tests can test for interactions but *t*-tests cannot.

6. Both F-tests and *t*-tests are robust, which means that many of the underlying assumptions for both tests can be violated with very little effect on their accuracy.

All the tests of significance, Z, Chi-Square (χ^2), *t*, and F, are basically related. [2]

Illustration 3: Relationships between Z, Chi-Square, t, and F

$$\sqrt{F} = t \approx Z \approx \sqrt{2x^2} - \sqrt{2(df) - 1}$$

$$F = \frac{x^2{}_1/df_1}{x^2{}_2/df_2}$$

$$t^2 = F$$

As one can see from inspection of the formulae, Z is a function of χ^2, and F is a ratio of $2\chi^2$ with their respective df.

Basically, all these tests have the same or very similar underlying assumptions. [3]

One can also see by inspection of the formulae for calculating *t* and Z, they are very similar. However, *t* distributions vary with the df. When the df for the *t*-test is small, its distribution is skewed. As the df becomes larger than 30, the distribution approximates the normal distribution of the Z. When the *t*'s df is over 400, the test and Z-test will produce, for all practical purposes, the same probability. Therefore, for a large df either Z or *t* can be used; but when the df is small, *t* should be used.

How to Use a Table to determine if r is Significant

Tables have also been developed to determine if r is significant for specific degrees of freedom. Table 7 is an example of this. The following is an example of how to use an r-table of significance.

Let's assume 32 people took both an achievement test and a creativity test (N = 32). We are interested in determining whether the correlation (either positive or negative) between the two tests was significant at the $\alpha = .01$ level. Since it doesn't matter if the correlation is positive or negative, it is non-directional (two-tailed test). The df = N - 2 = 32 - 2 = 30. You would then enter the table at df = 30 and $\alpha = .01$ for a two-tailed test to obtain a value of .45. If the calculated r is as large as or larger than .45, it means there is a significant relationship between intelligence and creativity at the .01 level.

Now let's assume we use the same example but do a directional (one-tailed) test. We hypothesize that there is a positive relationship between achievement and creativity. This time, we enter the table at the same degrees of freedom, but at $\alpha = .01$ for a one-tailed test, and obtain a value of .41. If the calculated r is as large as or larger than the table value of .41, it means there is a significant positive relationship between the tests. If the relationship found is negative, we must state that it is not significant since it is not in the direction we hypothesized.

Table 7: Example of a Table for the Level of Significance of r

df	One Tail (predicting the direction of the relationship)			
	.05		.01	.005
	Two Tail (not predicting the direction of the relationship)			
	.10	.05	.02	.01
1	.98	.997	.99	.999
5	.67	.76	.84	.87
10	.50	.58	.66	.71
20	.36	.43	.50	.54
30	.30	.35	.41	.45
60	.22	.25	.30	.33
100	.17	.20	.23	.26

Note: The df for r is generally N-2 where N is equal to the number of pairs of subjects in the study. Also note the values in the table are approximate.

Chapter Summary

Chapter III has dealt with statistical significance. Merely stating that something is significant does not provide us with enough information to intelligently draw conclusions. We must also be given the probability level, sample size, mean, and standard deviation.

The probability level tells us the probability that the relationship is not likely to be due to chance. We also know that as the sample size increases, the likelihood of finding significance also increases even though the magnitude of the relationship (magnitude of the effect) has not changed. The effect's magnitude or relationship is indicated by the mean score and standard deviation.

Statistical significance is operationally defined by the alpha level one chooses. This is a subjective decision. However, in the fields of education and psychology one usually selects .05, .01, or

.001. The alpha (α level) is the subjective probability statement of how often one is willing to accept something occurring by chance and still regard it as a non-chance occurrence.

The probability of making a Type I error is the same as the alpha level selected. It is the probability of saying a relationship exists when no relationship really exists. It is the probability of falsely rejecting the Null Hypothesis, which states there is no relationship. A Type II error is failing to reject the Null Hypothesis when it should be rejected.

When testing for significance, one a priori decides on making a one- or two-tailed test. A one-tailed test is a directional test and should be used only when you have prior information indicating the direction of the results. It is a more powerful test when used appropriately. A two-tailed test is non-directional and should be used when no prior information exists.

The most commonly used tests of significance are the Z, *t*, χ^2, and F. They are all functionally related. The *t* and F-tests are very robust. Many of their underlying assumptions can be violated without affecting their results. The *t*-test is a special case of more general F-test. Whatever you can do with a *t* can be done with a F-test, but the F-test is capable of doing more.

These tests of significance are used for calculating the probability of a relationship existing between variables. The values of these tests of significance are then generally looked up in tables to determine their probabilities. These probabilities are then compared to what one has previously decided to accept as a subjective alpha level. If the probabilities are equal to or less than ($p \leq \alpha$) the alpha decided upon, we state the relationship was found to be significant at that specified alpha level. In other words, the alpha level is our criterion for accepting something as being significant, and the probability (p) is calculated to see if it meets our criterion. If it does, we say it is significant; and if it does not, we say it is not significant.

Examples of how to read and use tables of significance were also presented. When using these tables, the calculated value must be as large or larger than the associated table value for a specified df and α level in order for the test to be significant.

---

[1]One can decrease the probability of making a Type I or II error by increasing the number of subjects used. This also increases the power of your test.

[2] Degrees of freedom (df) is a function of the number of independent replicates you have in your subject which is similar to the number of subjects in the study.

[3] The *t* and F-tests are very robust. Edwards (1972) states that these tests are practically insensitive to violations of the assumptions of normality of distributions and heterogeneity of variance (unequal variance in your groups) as long as the number (n) of subjects in each group are equal ($n1 = n2$, etc.) and $n \geq 25$.

Objectives for Chapter IV

After completing Chapter IV, you should be able to:

1. Graph data presented in tabular form.
2. Identify the area of interest on a graph.
3. Define interaction.
4. Define disordinal and ordinal interaction.
5. Identify main effects and simple effects on a given set of data.
6. Explain when it is appropriate to interpret main effects and simple effects.
7. Explain when it is appropriate to graph data using a solid line and when to use a dotted line.

Chapter IV

Introduction to the Concept of Interaction

In reading through any statistical literature one often encounters statements concerning interaction, i.e., "there was a significant interaction." Understanding this concept is frequently difficult, especially since it is usually introduced using a traditional statistical procedure as an explanation. The purpose of this chapter is to introduce the basic concept of interaction, conceptually and graphically, not mathematically. Our explanation should enable you to better interpret and understand the meaning of statistically significant interaction.

Definition of Interaction

The definition of interaction that we will use is: interaction exists when there is a significant differential effect across the areas of interest. In other words, when we plot the mean scores for at least two groups, with two or more variables, the relationship between these variables is seen by lines that are significantly not parallel.

Example of Plotting Interaction Data

Assume that we have one variable, the treatment (in this case the method of conducting reading), with two levels (two kinds), which we label treatment 1 and treatment 2. We also have another variable, teacher gender, with two levels, male and female. This gives us four groups, treatment 1 with a male teacher, treatment 2 with a male teacher, treatment 1 with a female teacher, and treatment 2 with a female teacher. If we look at the mean scores for each of the groups, we get a 2 x 2 table, as seen below.

Table 8: Example of a 2x2 Table

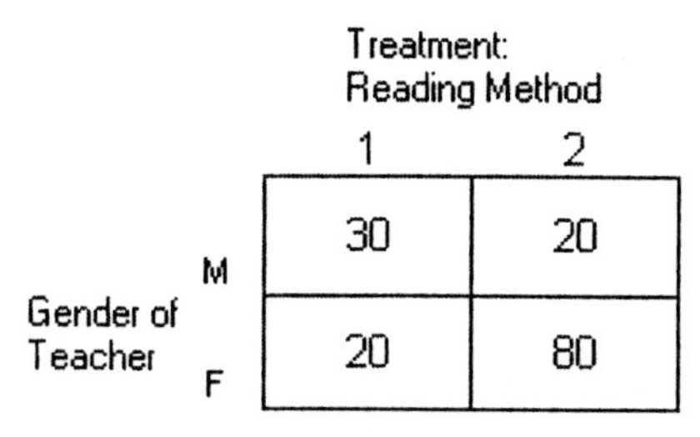

Gender of Teacher	Treatment: Reading Method 1	Treatment: Reading Method 2
M	30	20
F	20	80

Assume the mean reading achievement score for Treatment 1, when there is a male teacher, is 30. The mean score for Treatment 2, with a male teacher, is 20. The mean score for a female teacher in Treatment 1 is 20, and for a female in Treatment 2 is 80. These four mean scores can be plotted so that they are presented graphically as in Figure 19. This way, we can see the relationship between the variables.

Here, the area of interest when discussing interaction in graphic form is whatever is labeled on the X-axis. In the above example, the area of interest is the gender of the teacher.

Figure 19: A Plotted Interaction

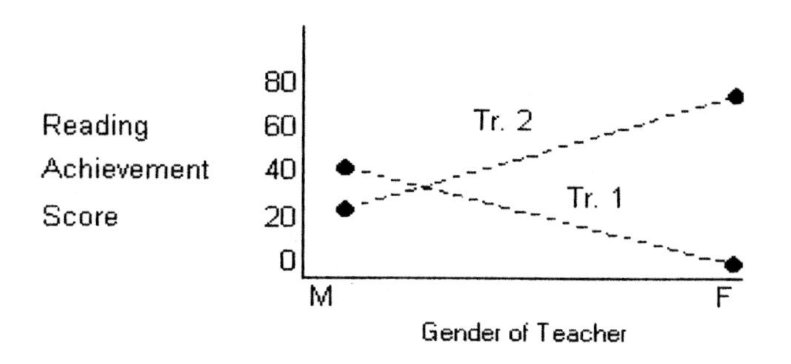

What Can be Interpreted When There is Significant Interaction?

Let's assume a *t*-test was conducted between Treatments 1 and 2 (the two methods of conducting reading), and it was found to be significant at the .05 level. In the above example, the mean score for Treatment 1 would be 25 {(30 + 20)/2}, and the mean score for Treatment 2 would be 50 {(20 + 80)/2}. Since Treatment 2 has significantly higher reading achievement scores, some may want to suggest that Treatment 2 be used in preference to Treatment 1. This would be an inappropriate conclusion since looking at the graph clearly indicates that Treatment 2 is only better when there was a female teacher. However, Treatment 1 was better when the teacher was male.

In the above example, there are two major variables (factors) also called *main effects*. One is treatment (1 and 2), and the other is gender of the teacher (male and female). If there is significant interaction, one should not, and technically cannot, interpret the main effects. That is, one cannot logically or technically say that Treatment 1 is better than Treatment 2. This is the reason why it is inappropriate to interpret the main effects of the treatment in the previous example.

As indicated, interaction may be conceived of as lines being significantly different from parallel when they are graphically plotted. In the above example, the plotted lines crossed. However, lines can also be significantly non-parallel without crossing. Both of the plotted lines below are examples of interactions.

We also stated previously that interaction is a differential effect across the area of interest and that the X-axis is the area of interest. In other words, on the two figures above, if there is a significant interaction, the difference between points A and B will be significantly different than the difference between C and D. Similarly, the difference between E and F will be significantly different than the difference between G and H. By definition, these lines are significantly non-parallel.

Figure 20A: Example of Plotted Lines

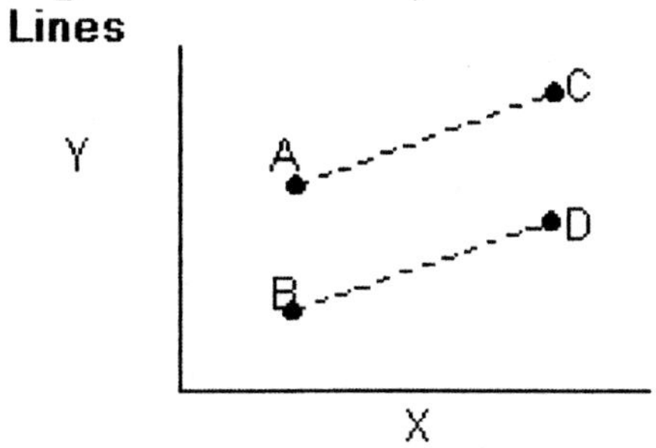

Figure 20B: Example of a Plotted Interaction

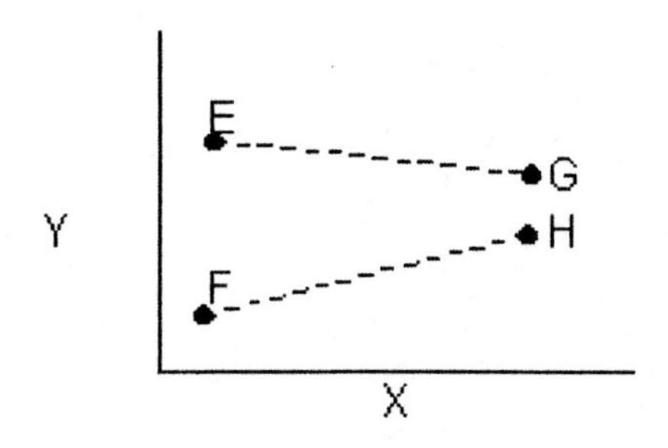

Two examples of parallel lines follow (Figures 21A and B).

Figures 21A&B: Examples of Plotted Main Effects With Parallel Lines, No Interaction

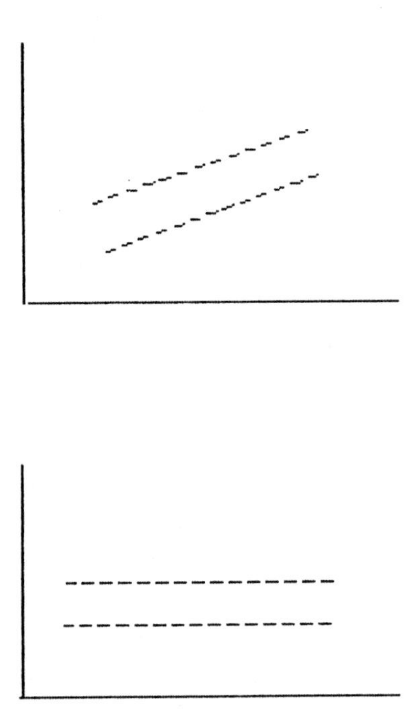

In graphs that resemble these, it would be appropriate to interpret main effects since there is no interaction.

Ordinal and Disordinal Interaction

There are two types of interaction, ordinal and disordinal. When the lines plotted cross within the values on the graph, the interaction is *disordinal.*

When the plotted lines do not cross within the values on the X-axis, this has been operationally defined as *ordinal interaction.*

Illustration 4:
Disordinal Interaction

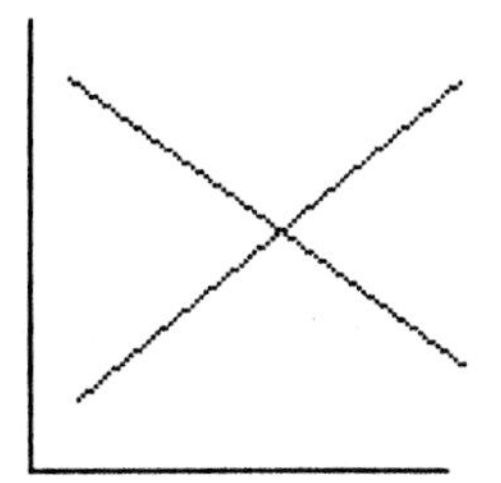

The first example presented in this chapter (Figure 19) was a case of disordinal interaction since the lines crossed, and Figures 20A and 20B were ordinal since the lines did not cross within the area of interest.

It was stated earlier that when there is interaction one can not interpret main effects. This is technically true. However, it may sometimes make logical sense to interpret main effects when there is ordinal interaction.

Illustration 5:
Ordinal Interaction

The more traditional statistical texts will state that only simple effects (looking at each group independently) must be interpreted when there is significant interaction. In our first example, there was significant interaction. There were four groups; therefore, to interpret the simple effects, one would look at each of these groups separately. The conclusions based on this interpretation would be that Treatment 2 is more effective when there are female teachers, and Treatment 1 may be more effective when there are male teachers (depending on whether or not the difference is significant).

Plotting Interaction for Three Treatments Across Gender

For another example, let's assume that there are three treatments and gender of the teacher.

Table 9: Example of a 2x3 Table

		Treatment		
		1	2	3
Gender of Teacher	M	65	75	90
	F	70	80	20

Let's also assume that the six groups mean scores on the reading achievement test are presented in the boxes in Table 9. That is, the mean score of Group 1, which is Treatment 1 with male teachers, is 65, and so forth for the other five groups. In one were to graph these means, they may appear similar to Figure 22.

Even though we have six groups in this case, we still only have two main effects, treatment and the gender of the teacher. As stated, if there is significant interaction, one cannot technically make statements about main effects but can only discuss simple effects (looking at each of the six groups separately).

As one can see by looking at Figure 22, even though Treatments 1 and 2 appear to be parallel (therefore not interacting with each other) they both interact with Treatment 3, over the area of interest (the gender of the teacher).

One possible interpretation of this data is that Treatment 3 may be the most desirable to use when the teacher is male, and it is potentially the worst treatment when there is a female teacher. Treatment 2, on the other hand, may be the most appropriate if both male and female teachers had to be used. These are examples of statements of simple effects since we cannot make a more general statement such as Treatment 1 is the best regardless of the teacher's gender, etc. In other words, in this example a decision as to which treatment is to be used must be based on the treatment and whether the teacher is male or female. These data are also an example of disordinal interaction, since the lines cross within the area of interest.

Figure 22: Example of Plotting for Three Treatments Where There is an Interaction

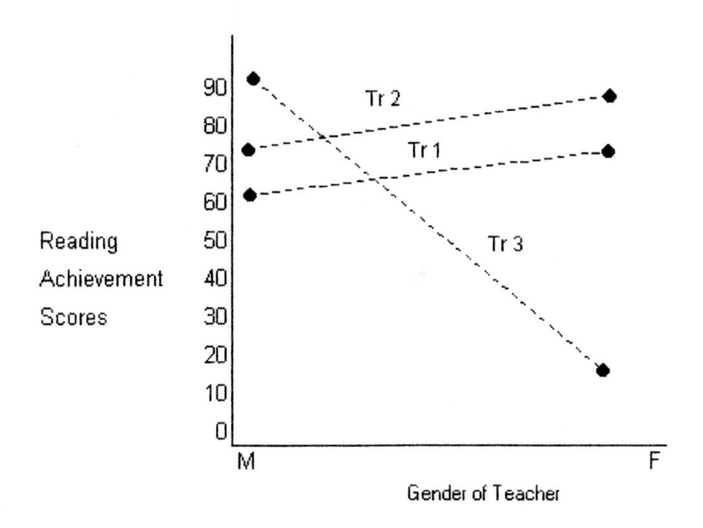

Dotted and Solid Lines

Up to this point we have used dotted lines to connect points on the graphs. The accepted symbolic representation is to use a dotted line when connecting points over an area of interest that is made up of *dichotomous or categorical variables.* In other words, a dotted line is used when the area of interest has a variable that is not at least ordinal in value. A solid line is used to connect points when the area of interest consists of at least *ordinal* data (it can be ordinal, interval, or ratio). Examples of when a solid line should be used are when time, income, age, years of experience, etc., are the variables on the X-axis area of interest.

If the data looked like the following, it should be graphically represented by using a solid line (see Table 10 and Figure 23).

Table 10: Example of a 2x3 Table for Treatments by Income Level

		Treatments 1	Treatments 2
Income	Low	60	10
	Medium	70	40
	High	90	70

Figure 23: Example of Plotting of An Ordinal Interaction

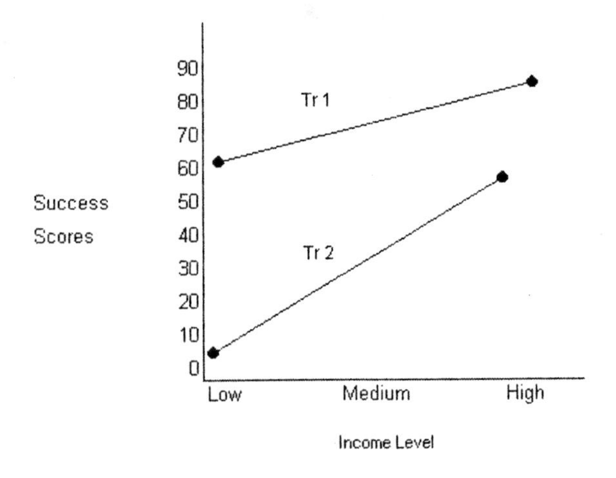

As one can see, the lines do not cross within the area of interest (achievement); therefore, if there is significant interaction, it would be an ordinal interaction.

Chapter Summary

In this chapter, interaction was defined as the differential effect across the area of interest (whatever is on the X-axis). In other words, the magnitude of the differences will be significantly different for various values across the X-axis and is depicted graphically as significantly non-parallel lines.

Disordinal interaction occurs when the lines cross within the area of interest. When they cross outside of the area of interest, the interaction is ordinal.

The variables on the X and Y axes are called the main effects. These are not technically interpretable when there is significant interaction. This is always the case with disordinal interaction; but sometimes, when the interaction is ordinal, it may be logically interpretable.

When interaction exists, it is always appropriate to interpret the simple effects. This is accomplished by looking at each group independently and comparing it to every other group.

To graphically plot data, we may use a solid or dotted line. A dotted line is used when the data are not at least ordinal in nature, and a solid line should be used when the data are at least ordinal.

Objectives for Chapter V

After completing Chapter V, you should be able to:

1. List the purposes of research design.
2. Interpret the meaning of a significant F when it is conducted on two or more groups.
3. Given any two-factorial design (2 x 2, 3 x 3, etc.), draw the cells and identify the main effect and simple effects.
4. Given an analysis of variance table and the total number of subjects, calculate the degree of freedom for the main effects, interaction, within, and total.
5. Interpret the F-test for factorial design.
6. Given a three-factorial design (2 x 2 x 2, 4 x 2 x 3, etc.), draw the cubes and identify the main and simple effects.
7. List the important considerations when using factorial design, as stated in this chapter.
8. Define and discuss the problem of intact groups.
9. Name the variable that is always statistically analyzed.
10. Determine the total df when given between groups df and within groups df.
11. Determine the between groups df when given the df for the main effects and interaction.
12. State the relationship between determining total df and total SS.
13. See hyperlinks.

Chapter V

Introduction to Factorial Design and Its' Interpretation

Most published research tends to use analysis of variance (the F-test) to analyze factorial designs. This is used so frequently that people have confused the two. Analysis of variance is a statistical procedure, and factorial design is a design.

Purpose of Research Design

The purpose of design is to control for certain variables while testing others. After the research questions have been decided upon, the researcher then designs his study to control variables that he thinks may be of concern or eliminate variables that he thinks may contaminate the study.

The following is a list of how research design helps the researcher:

1. It controls variance.

2. It sets up the variables in such a way that the relationship between them can be adequately tested.

3. By analyzing the design, one can determine which statements can legitimately be made and the limitations.

Generally, design can be taught in two ways. One method is the Campbell and Stanley procedure. This stresses the conclusions that can be made from the analysis of the design, the limitations, and the population to whom the results can or cannot be generalized. This method also stresses the logical alternative explanations that may have caused the difference, other than the variables controlled for by the design. The Campbell and Stanley method does not stress the statistical procedures necessary to analyze the design since there are potentially many appropriate procedures.

The second method is factorial design. This method has its greatest stress on the statistical procedures for analyzing the design. This chapter will briefly discuss this second procedure to the extent that the reader will be able to interpret it when a factorial design is seen in the literature. We will not go into the statistical calculations.

One-way Analysis of Variance Design

The simplest design is a one-way analysis of variance design. The simplest of these is testing to see if there is a difference between two groups; for example, group 1 and group 2. In this design, Treatment is the only factor, and it is therefore not a factorial design because factorial designs must have at least two factors.

Figure 24: Example of a One-Way Analysis of Variance Design

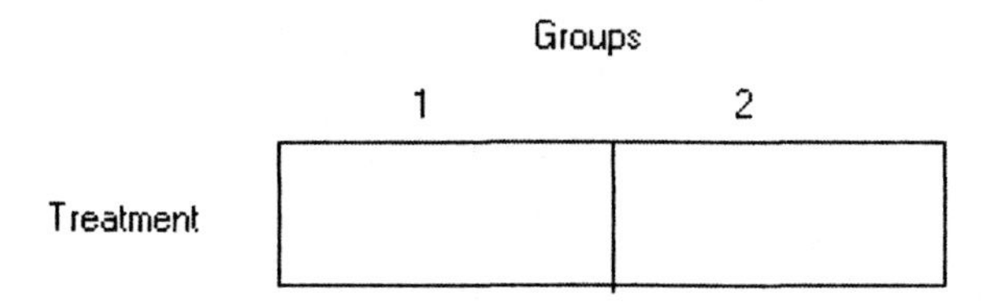

This is similar to a *t*-test between two groups. As was pointed out in Chapter 3, when you have only two groups, the F-test is actually equal to t^2; and you get the same results.

Another analysis of variance example, in which the F-test is more appropriate than the *t*-test because there are more than two groups, is when there are five groups (see Figure 25).

Figure 25: Example of a One-Way Analysis of Variance Design with Five Groups

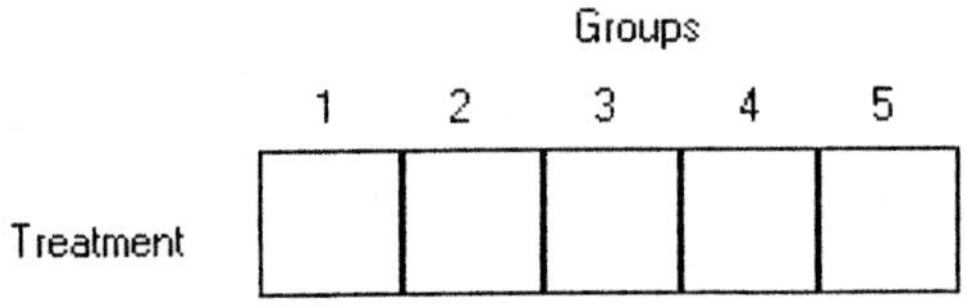

	Groups				
	1	2	3	4	5
Treatment					

If we ran an F-test on the five groups and found it to be significantly different, we would not know what caused the difference only that some group or combination of groups was significantly different. Quite often, when a significant F is found, some researchers will go back with a *t*-test or F-test and test for the significant difference between specific groups to locate the difference. (There are more technically accurate procedures to do this, but they will not be discussed here.)

The following link will perform the calculation of a one-way analysis of variance (with up to 5 groups being compared): http://faculty.vassar.edu/lowry/anova1u.html.

Example of a 2 x 2 Factorial Design

The simplest example of a factorial design is a 2 x 2 design which must have at least two factors, and each factor must have at least two levels. In other words, the simplest factorial design will produce at least four groups (2 x 2 = 4). As stated in Chapter 4, this is the minimum number of groups needed to test for interaction (see Table 11 for an example of this design.)

Table 11: Example of a 2x2 Factorial Design Table

		Gender	
		Male	Female
Treatment	1	40	60
	2	80	55

In this example, we have two factors, gender and treatment. Gender has two levels, male and female, and treatment has two levels, 1 and 2.

Illustration 6: Breakdown of a 2 x 2 Factorial Design

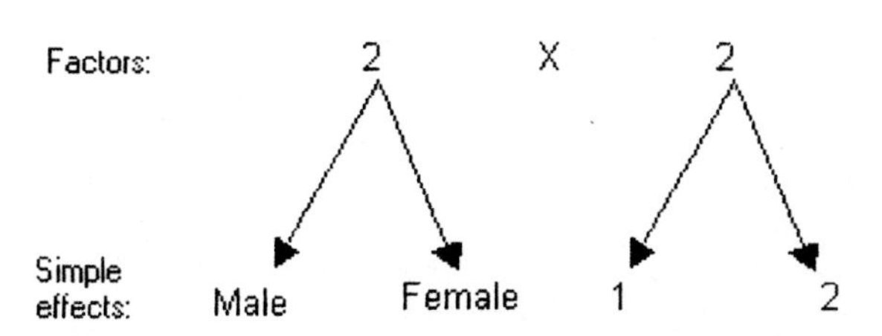

The factor gender is considered one main effect, and the factor treatment is the second main effect. These two main effects with their two levels create four groups that are:

Treatment 1 – male
Treatment 1 – female
Treatment 2 – male
Treatment 2 – female

When one looks at the separate groups, they are called the *simple effects*.

Quite often when one presents the analysis for a factorial design in table form, the effects are sometimes represented by letters. For example, gender may be represented by the letter A, treatment by the letter B, and the interaction by A x B (gender x treatment). The table that presents these results generally takes the form of Table 12.

To interpret Table 12, you really only need the *source, df, and F*. As stated in Chapter 3, to determine if an F is significant there must be two sets of degrees of freedom, the degrees of freedom numerator (df_N), and the degrees of freedom denominator (df_D). Since there are three F values in this particular table, two sets of degrees of freedom are needed for each F. In the general cases we are dealing with, the df_D will always be the df associated with the within (error), which is 100 on this table. The second df, the df_N is the degrees of freedom associated with each of the sources of variability listed in the column source.

Table 12: Example of Analysis Table for a 2x2 Factorial Design

Source	df	SS	MS	F
A (sex)	1			1.5
B (treatment)	1			10.2
A x B	1			12.2
Within (error)	100			
Total	103			

In this table, there are three sources of variability that are controlled for; the sources of variability due to gender, the source of variability due to treatment, and the source of variability due to the interaction between gender and treatment.

Calculating Degrees of Freedom

The degrees of freedom are calculated in the following manner:

For the A main effect: df = a – 1
Where: a = number of levels of A (gender), in this case a = 2
df for A = a –1 = 2 – 1 = 1

For the B main effect: df = b – 1
Where: b = number of levels of B (treatment), in this case b = 2
df for B = b – 1 = 2 – 1 = 1

For the A x B interaction: df = (a – 1) (b – 1) df for A x B = (2 – 1)
df for A x B = (2 – 1) (2 – 1) = (1) (1) =1

For the within (error): df = N – K
Where: N = total number of subjects in this study (we assumed N = 104)
K = number of groups
df for within = 104 – 4 = 100

Total df: can be obtained by adding up all the others or by N – 1.

If one was interested in testing to see if there is a significant effect due to gender, one would look up the F value in the table using the df associated with gender and within. In this case, if it the alpha level was at .05 (α = .05), it would be found to be not significant. In other words, gender did not account for a significant amount of variance. In checking to see if there was a significant difference due to treatment, one would enter the F-table using the df associated with treatment and within. At an alpha of .05 this would be significant indicating that treatment accounted for a significant amount of variability. Following the same procedure for interaction, one would use the df values for the within and interaction. This would also be found significant at the α = .05 level.

If you recall from Chapter 4, when significant interaction exists, it is technically not appropriate to interpret the main effects even if they are found to be significant, as treatment is in our example. If there is significant interaction, it should be graphed, as demonstrated in Chapter 4, to determine if the interaction is ordinal or disordinal. If we assume that the dependent variable being measured is a reading achievement score, which would be graphed on the Y-axis, gender may be considered the area of interest and would be graphed on the X-axis, as in Figure 26. If the scores obtained were like the ones in our example, then the interaction would be disordinal, and no interpretation of main effects can be made.

Figure 26: Graph of a Disordinal Interaction

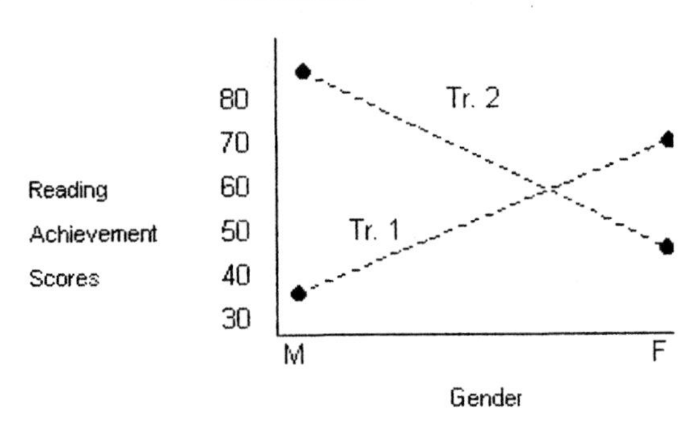

Figure 27: Example of a 2x3 Factorial Design

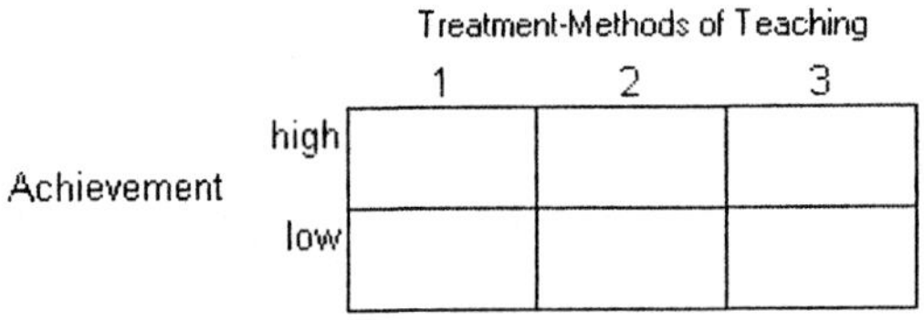

Achievement	Treatment-Methods of Teaching 1	2	3
high			
low			

Example of a 2 x 3 Factorial Design

In a 2 x 3 design, we have two factors, one of which has two levels and the other has three levels. There are six groups (2 x 3 = 6). Let's assume the first factor is achievement and it has two levels, high and low. The second factor is treatment, with three methods of teaching arithmetic. (see Figure 27 as an example of a 2 x 3 factorial design).

The traditional analysis of variance table will look like the example in Table 13, which assumes that the total number of subjects is 100.

If the F-ratios were looked up in a table, one would find the interaction is not significant at the α = .05 level. (The df_N would equal 2 and df_D would equal 94.) Since the interaction is not significant and the main effects are, it is legitimate to interpret main effects *.

The A main effect (achievement) only has two groups, therefore, one can easily interpret this to mean that there is a significant difference between high and low achievement. However, no directionality is given. To determine which group scored higher on the dependent variable (criterion), one must look at the mean scores for the high and low achievement groups.

Table 13: Example of a Traditional Analysis of Variance Table

Source	df	SS	MS	F
A (achievement)	(a-1 = 2-1) = 1			4.1 (S)
B (treatment)	(b-1 = 3-1) =2			6.3 (S)
A x B	(a-1)(b-1) = (1 x 2) = 2			.2 (NS)
Within (error)	(N - k = 100 -6) = 94			
Total	(N-1) = 99)			

Note: a = number of levels of the A main effect (factor 1)
b = number of levels of the B main effect (factor 2)
N = total number of subjects in the sample, in this case we assumed 100 subjects
K = total number of groups (2 x 3 = 6)
S = significant, ns = nonsignificant

The B main effect was also found to be significant; but since there are three groups, the interpretation of the significant F just means that there is a difference somewhere in the three groups (the same interpretation as if we were doing a one-way analysis of variance on three groups). The following link will perform the calculation of a two-way analysis of variance (with up to 4 levels per independent variable – therefore up to 4x4 design): http://faculty.vassar.edu/lowry/ anova2u.html.

Example of a 2 x 2 x 2 Factorial Design

Here we have three factors, each with two levels. Let's assume factor A = treatments, Factor B = gender, and Factor C = achievement. This would be depicted as a three-dimensional figure (a cube) (See Figure 28).

In a 2 x 2 x 2 factorial design, there eight groups (2 x 2 x 2 = 8) and three factors (three main effects). The eight groups are:

Figure 28: Example of a 2x2x2 Factorial Design

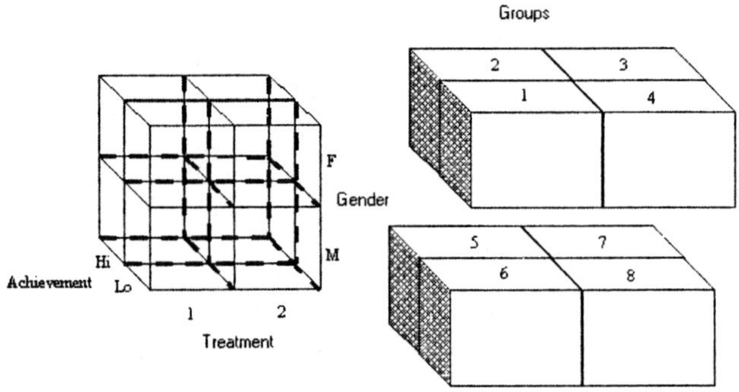

Group 1 = females with low achievement in Treatment 1
Group 2 = females with high achievement in Treatment 1
Group 3 = females with high achievement in Treatment 2
Group 4 = females with low achievement in Treatment 2

Group 5 = males with high achievement in Treatment 1
Group 6 = males with low achievement in Treatment 1
Group 7 = males with low achievement in Treatment 2
Group 8 = males with high achievement in Treatment 2

These groups make up the simple effects.

The analysis of variance table for this is somewhere more complex since there are three main effects that allow for three two-way interactions and one three-way interaction. This is illustrated in Table 14.

Please see Appendix T for a brief discussion and an example of a correction for multiple comparisons.

The significance for the main effects and interactions is looked up in the F-table in exactly the same way as in the previous example. You need two sets of df for each. In this case, the df_D will always be 192, and the df_N for each one of these is one. If any of the two-way interactions was found to be significant, then the main effects could not technically be discussed.

Table 14: Analysis of Variance Table for a 2x2x2 Factorial Design

Source	df	SS	MS	F
A (treatment)	(a-1) = 2-1 = 1			7.2 (s)
B (sex)	(b-1) = 2-1 = 1			1.5 (ns)
C (Achievement)	(c-1) = 2-1 = 1			4.5 (s)
A x B	(a-1) (b-1) = (2-1) (2-1) = 1			.7 (ns)
A x C	(a-1) (c-1) = (2-1) (2-1) = 2			1.2 (ns)
B x C	(b-1) (c-1) = (2-1) (2-1) = 1			.8 (ns)
A x B x C	(a-1) (b-1) (c-1) = (2-1) (2-1) (2-1) = 1			1.1 (ns)
With (error)	(N-K) = (200-8) = 192			
Total	N-1 = 199			

Note: a = levels of the A main effect (2)
b = levels of the B main effect (2)
c = levels of the C main effect (2)
K = total number of groups (2x2x2=8); in this example there are 8 groups
N = total number of subjects (assumed for this example to be 200)
s = significant; ns = non-significant

However, in this case, there is also a three-way interaction, which if found to be significant, would mean that the two-way interactions could not be discussed. In this event, each of the eight groups would have to be discussed and compared separately.

The following link will perform the calculation of a 2x2x2 analysis of variance: http://faculty.vassar.edu/lowry/an2x2x2.html

Important Considerations

1. In the beginning of this chapter, we stated that the purpose of research design is to control variance. The more variables (*factors*) we have in a design the more variance we can control.
2. The more factors and levels of factors one has, the more groups one has and the more subjects one needs.
3. One rule of thumb when working with factorial design is never to have less than five subjects in a group, even though the analysis can be run with two subjects. It is desirable to have at least twenty subjects per group.
4. There are many assumptions that underlie analysis of variance that we will not deal with. However, it is important to mention that the F-test is very robust and does not have to worry about most of the assumptions especially if two conditions exist. These are that there are at least twenty subjects in each group (the more the better) and that all groups have the same number of subjects.
5. One assumption that must be met is that of independence of measurement. This means that each subject's response on the *dependent variable (criterion)* is independent of any other subject's response. For example, suppose we had two groups (ten people each) of individuals in group therapy. Group 1 received Rogerian Therapy and Group 2 received Gestalt. N, in this case, would not be twenty. It would be two since each subject's success in therapy is not independent of the other members in the group. The only things that are independent are the groups. This has been referred to as the problem of *intact groups*. It occurs very frequently in the field of educational research. For another example, when one is testing to determine which treatment is more effective, four different classes of thirty students in a class may be used. The N would be 4 instead of 120. This could make a dramatic difference in determining whether or not there is statistical significance.

6. Only the dependent variable (criterion) gets statistically analyzed. It is the score that is used to compare each of the groups (simple effects and main effects).
7. The between subjects and the within subjects = Total Subjects. Similarly, the between df plus the within df = total df. In a factorial design (for example 2 x 2), the between df is equal to the df of the A main effect plus the df of the B main effect plus the df of A x B interaction. Therefore, total df = between df plus within df (between df = Adf + Bdf + A x Bdf) and (total df = between df + within df).
8. In a three-dimensional factorial design (for example 2 x 2 x 2), the between df = df for A main effects plus the df for B main effects plus the df for C main effects plus the df for A x B plus the df for A x C plus the df for B x C plus the df for A x B x C. Total df = between df plus within df. This same principle applies to the subjects.

Chapter Summary

Chapter 5 has dealt with the basic interpretation of factorial design. The purpose of research design is to control for variability (variance). This allows for adequately testing relationships and enables us to determine appropriate conclusions and the limitations.

The simplest design is a one-way analysis of variance design in which there is only one factor that may have any number levels. The simplest factorial design is two-factorial design (2 x 2) having two factors (main effects) of two levels each, creating four groups (simple effects). A three-factorial design has three factors (three main effects). The number of simple effects is determined by the levels of each factor (2 x 2 x 2 = 8; 2 x 3 x 2 = 12; 6 x 3 x 4 = 72). In the 6 x 3 x 4, the A main effect has six levels, the B main effect has three levels, and the C main effect has four levels.

To determine if an F is significant, two sets of degrees of freedom are needed. The degrees of freedom-numerator (df_N) is always the df associated with that particular main effect. The degrees of freedom-denominator (df_D) is always the df associated with the within. The between df plus the within df equals the total df.

If there is a significant interaction, one can only interpret the simple effects in a two-factorial design. When dealing with a three-factorial design, one cannot interpret the main effects if any of the two-way interactions is significant. Similarly, one cannot interpret any of the two-way interactions if there is significant three-way interaction. A significant three-way interaction means there is a significant interaction between at least two, two-way interactions.

To avoid worrying about the underlying assumptions, one should have as large an N as possible and keep equal numbers of subjects in each group. As a rule of thumb, the N should not be less than five subjects per group.

A problem frequently encountered in education research is one of intact groups. The N is always the number of independent replications. It is not necessarily the number of people involved. One should also keep in mind that only the dependent variable (criterion) gets statistically analyzed.

---

* The df for determining the significance of the A main effect are 1 and 94 and for the B main effect, the df are 2 and 94.

Objectives for Chapter VI

After completing Chapter VI, you should be able to:

1. Identify and define:
   - A. pre-experimental designs
   - B. three types of ex post facto research
   - C. quasi-experimental designs
   - D. true-experimental designs
2. Discuss internal and external validity as they relate to each of the designs in Objective 1, a-d.
3. Identify and define internal validity.
4. Identify and define external validity.
5. Identify, define, and give examples of independent variables (active variables) that can be manipulated.
6. Identify, define, and give examples of assigned or attribute variables that cannot be manipulated.

Chapter VI

Introduction to Research Design: Internal and External Validity*

Ideally, the purpose of research design is to control for all possible alternative explanations other than the one being investigated. To the extent that the research design can do this, the design is *internally valid;* and to that extent, one can infer causal relationships. Designs with total internal validity are referred to as "true experimental designs."

True experimental designs require so much control that they generally can only be conducted in a laboratory setting. Therefore, when you are doing research in an applied setting, the controls you need for true experimental designs are generally lacking so other designs must be used. These alternatives are generally classified as pre-experimental, quasi-experimental and ex post facto designs.

Pre-Experimental Designs

Pre-experimental designs have virtually no internal validity, even though they have an independent variable that is capable of being manipulated. An example of this type of design might be pretest-treatment-posttest (O X O).

This common design is generally used to infer that the treatment caused the change in scores from the pretest to the posttest. However, many other explanations are possible, such as the pretest may have sensitized the subject which improved posttest scores or that time lag between the pre- and posttest may have caused the difference. Another example of a common pre-experimental design is one in which a treatment is given to one of two groups and the groups are tested for differences.

This is symbolized by $\frac{X\ O}{O}$. However, there is no way of knowing if the treatment caused the difference or if the two groups were different to begin with since they were not initially tested for equivalence. As one can see, pre-experimental designs do not have very much control even though they have an independent variable that can be manipulated by the researcher (an active variable).

Quasi-Experimental Designs

Quasi-experimental designs have a little more control and have independent variables that are under the control of the experimenter. However, they do not have enough control to be considered true experimental designs. Generally, in quasi-experimental designs, one does not have enough control over the situation to randomly assign subjects to treatments or to have control over the scheduling of the testing or treatment. However, some of the quasi-experimental designs may range from very little to a great deal of internal validity depending upon the specific situation.

Ex Post Facto Research

Ex post facto research is generally a term that describes research which is initiated after the independent variable (the variable of interest) has already occurred or the independent variable is a type that cannot be manipulated such as age, race, gender, economic status, etc. Ex post facto research has been subject to more criticism by research methodologists than other research. Some of it is justifiable, and some is not. This type of research is often done in education. More on this research is found later in the text.

The following table represents the theoretical relationships between pre-experimental, quasi-experimental, ex post facto and true experimental designs in terms of internal validity (see Table 15).

Table 15: The Continuum of Internal Validity as Related to the Various Possible Research Designs

	1	2	3	4	5	6
	Lowest					Highest
CRITERIA WITH WHICH TO JUDGE INTERNAL VALIDITY	Pre-Experimental Designs	Ex Post Facto with No Hypothesis	Ex Post Factor with Hypothesis	Ex Post Facto with Hypothesis & Tests of Alternative Hypothesis	Quasi-Experi-mental Designs	TRUE Experi-mental Designs
1. Random assignment occurs within design:	NO	NO	NO	NO	NOT GENERALLY	YES
2. Definite controls for scheduling:	NO	NO	NO	NO	QUESTIONABLE YES/NO	YES
3. Independent variable is active and can be manipulated by the experimenter:	YES	NO	NO	NO	YES	YES
4. Independent variable is assigned/attribute and cannot be manipulated by the experimenter	NO	YES	YES	YES	NO	NO

Klein's Modification of Newman's 1976 Continuum of Research Design.

Pre-experimental and ex post facto designs with no hypotheses are generally considered to be the weakest in terms of internal validity, that is, no causal inferences should be made. Ex post facto research with hypotheses has potentially much more scientific value. However, ex post facto research with hypotheses and tests for alternative hypotheses is considerably more powerful in terms of internal validity than the preceding designs, and it even may be better than some types of quasi-experimental designs, depending on the strength of the controls. Obviously, true experimental design is the most powerful in terms of internal validity.

External validity is defined as the ability to generalize results from the testing situation to the general population that was not tested. In research, one finds that as the experimental controls are increased, one's ability to generalize beyond the controlled testing situation is decreased. Therefore, the most controlled designs with the greatest internal validity tend to have the least external validity. So, if one looked at the experimental, quasi-experimental, and ex post facto research, true experimental research would have the least amount of external validity while ex post facto research would have the most. Following is a more detailed discussion of ex post facto research.

Throughout the literature one can find ex post facto research almost delegated to an inferior position among the types of research designs and methodology. The terms "ex post facto research" and "correlation research" are sometimes used interchangeably. When one does correlational (ex post facto) research, causation cannot be inferred. Therefore, many methodologists have issued numerous warnings emphasizing the dangers and possible misinterpretations of research in which the experimenter does not have control over the independent variables.

In ex post facto research, causation is sometimes improperly inferred because some people have a propensity for assuming that one variable is likely to be the cause of another because it precedes it in occurrence, or because one variable tends to be highly correlated with another (for example: Smoking—the independent variable, assumed to cause cancer—the dependent variable). This obviously does not necessarily mean because two variables are correlated and one precedes the other that they are not causally related. However, while a correlated and preceding relationship is necessary, it is not sufficient for inferring a causal relationship.

To assume a causal relationship, one must have internal validity (all other explanations for the effect on the criterion [dependent variable] are controlled for and the only possible explanation for changes in the dependent variable must be due to the independent variable under investigation). Only with a true experimental design does one have the experimental control to achieve internal validity. Ex post facto research lacks this control for a variety of reasons. One of these reasons is that there is an inability to randomly assign and manipulate the independent variable since in this research it has already occurred and is not under the control of the researcher.

A common weakness of ex post facto research is that the design is not capable of controlling the confounding effects of self selection. For example, suppose research was conducted see what effect early childhood training had on motivation; also suppose a significant relationship between early independence training and later adult motivation was found. Therefore, one might incorrectly conclude that the independence training "caused" this adult motivation. Another explanation might be that volunteer subjects who had independence training were more likely to demonstrate a greater degree of adult motivation but what causes this motivation might be more related to what causes them to volunteer than it is to the independence training.

A better example may be data that indicates elementary school males score more significantly lower on reading achievement than females; therefore, one may tend to assume that males are less competent in decoding this visual stimulus. One of many other reasons for their lower scores may be that in our culture, reading may be considered to be more of a feminine activity and, therefore, males may receive less reinforcement for engaging in this activity and may even receive aversive comments from their peers, which could affect reading performance.

Obviously, one can go on with examples of ex post facto research that can not appropriately assume causal relationships. Because of this well acknowledged weakness, many tend to regard ex post facto as inferior research that should not be conducted.

This is not necessarily true, since it is the research question that determines whether or not its use is appropriate. If the research question is dealing with causation, ex post facto procedures are inappropriate. However, if the question deals with relationships, it may be very appropriate.

Sometimes a research question of interest has independent variables that cannot be manipulated. Therefore, the researcher can either decide to do ex post facto research or no research at all.

The authors feel that, before deciding to conduct research, one should ask the question, "So what?" "So what if significance or non-significance is found?" Is this useful information? When the "So what?" question is answered positively, the researcher may find that the variables are the type that require ex post facto research. The research that tends to be related to some of the most significant social problems, by its very nature, has to be ex post facto (Kerlinger, 1972).

One of the most effective ways of using ex post facto research is to help identify a small set of variables from a large set of variables that is related to the dependent variable, for future experimental manipulation.

One can identify three major types of ex post facto research: *without hypotheses, with hypotheses,* and *with hypotheses and tests of alternative hypotheses.* Ex post facto with hypotheses tests previously stated hypothetical relationships. This type of research is much better; but there is still a danger of misinterpretation, and one must be cautious once again in interpreting the results of the investigation.

The third type of ex post facto research tests the stated hypotheses and alternative hypotheses. These are hypotheses that purpose other explanations for the effect, other than the stated ones. These explanations are competing or rival hypotheses to the ones the researcher is interested in verifying. The more of these rival hypotheses that can be eliminated, the greater the internal validity of the design. However, one must still keep in mind that by its very nature ex post facto research can never have total internal validity. Therefore, causation can never be inferred.

Chapter Summary

In this chapter, the purpose of research design is defined as controlling for all possible alternative explanations. A design is internally valid and causal relationships can be inferred to the extent that the design controls for these alternatives. A design is externally valid to the extent the results can be generalized beyond the testing situation. Therefore, the most controlled designs tend to have the least external validity.

These are generally four classifications of research designs. There are:

1. *Pre-experimental* – very little control, virtually no internal validity; therefore, no causal inference should be made.

2. *Quasi-experimental* – a little more control, but not enough to randomly assign subjects to treatments or to control scheduling of treatments. Internal validity ranges from very little to a great deal, depending on the situation.

3. *Ex post facto* – the independent variable has already occurred or cannot be manipulated. The design may be very weak in internal validity or relatively strong depending on whether there are hypotheses and tests for alternative hypotheses. They tend to have the most external validity.

4. *True experimental* – regarded as totally internally valid. One can infer causation, but these designs require so much control that a laboratory setting is necessary. Therefore,

they are not practical for most situations one wishes to investigate. They tend to have the most internal validity.

It was also stressed that causation can only be assumed when all other explanations for the effect are controlled for and the only possible explanation for change must be due to the independent variable under investigation. If the research question is concerned with finding a causal relationship, then ex post facto procedures are not appropriate.

Finally, researchers should be concerned with the "So what?" question. They should be concerned with whether the information produced will be useful, and they should be able to state why their research question is worth investigating before they invest their time, money, and labor.

---

* This presentation has been heavily based upon the work of Campbell and Stanley's classical monograph, Experimental and Quasi-Experimental Designs for Research (1996).

Objectives for Chapter VII

After completing Chapter VII, you should be able to:

1. Identify some of the common symbols used in diagramming research designs.
2. Define:
    a. A one shot case study
    b. A pre-test/post-test case study
    c. A simulated before-after design
    d. A two group-post-test only, nonequivalent design
    e. A pre-test/post-test nonequivalent design (a quasi-experimental design)
    f. A longitudinal time series design
    g. A post-test only experimental and control groups randomized design
    h. An experimental and control group matched subjects design
    i. Before and after true experimental design with a control group
    j. A Solomon 4-group design
3. For each of the above designs, be able to discuss the internal validity.

Chapter VII

Further Discussions on Research Designs

It is convenient to represent research designs with a diagram. The diagrams that are used to represent designs are called paradigms. Studying a paradigm makes it easy to plan, interpret, and analyze research.

In this chapter, ten designs will be presented. These ten represent the more common designs that are used in educational and behavioral science research. There are many more designs that could be presented; however, many of these additional designs are modifications of the basic ten presented in this chapter. Thus, the reader who understands these fundamental designs will have the basic tools for interpreting most research.

In representing research designs with paradigms, symbols are used that must be explained to the reader. The following is a list of symbols used in this chapter and a brief description of the meaning of each:

X this refers to the experimenter's treatment for a group.

-X no treatment or an absence of treatment.

In an experiment, one group might receive a treatment and a second group not. The second group is commonly called the control group.

Ⓧ that the independent variable is not manipulated; it is either an attribute or it is assigned.

O a measurement or an observation. In educational research, O will frequently be some type of test score. For example, pre-test and post-test. It can have any number of subscripts.

R means the random assignment of subjects to groups.

M means the assignment of subjects to groups using a matching procedure.

M_r means the assignment of subjects to groups by first matching subjects and then randomly assigning each matched subject.

The first group of designs presented (1-6) is quite common but these designs are somewhat inadequate in terms of internal validity. There are often one or more factors that could impact the research, and serve as alternative explanations of the possible causes of change. The following are the most frequent factors, borrowing terms from Campbell and Stanley's classical work, *Experimental and Quasi-experimental Designs for Research.*

History could account for the change in a group. History means that any extraneous event other than the treatment that intervened between the pre- and post measures. For example, if a school district was investigating the effect of a new set of arithmetic textbooks on achievement by measuring achievement at the beginning and end of the year, Design #2 (presented below) might be chosen. But what if during the year all the students moved into a new building? This would be an example of history factor that might affect the results of the study. In other words, the researcher wouldn't know if it were the new books or the new school that caused the results of the study.

Another example might be a researcher investigating the effects of integration on student attitudes toward people of different races. In our example, the history factor is the death of Martin Luther King one week before questionnaires were distributed to students. Obviously, such an event would drastically affect student responses and confound the results in an unknown way.

Maturation is another factor that could account for a change in a group. Maturation means any growth or development that normally takes place independent of an experimental treatment.

An example might be the analysis of the impact of a school lunch program on a student's growth. Part of the change that occurs in a student's growth would normally occur as part of their maturation. Often, history and maturation are confused because they both are related to something occurring in the time between the pre- and post-tests. The distinguishing point to keep in mind that *historical effects* are happening *external to the subject* while *maturation* effects are related to *changes internal to the subject.* Examples of these internal changes may be psychological such as boredom or physiological such as fatigue.

Testing refers to factors associated with measuring devices that actually cause change to occur. This may happen when a pre-test sensitizes people to an experimental treatment and actually causes them to behave differently during the treatment. An example of this would be a teacher's evaluation of the impact of a Values Clarification Program on the moral development of students. If the teacher gave a pre-test that asked questions about morality, the pre-test might possibly get the students concerned about the topic and thus influence them to be unusually receptive to the Values Clarification Program. This same phenomenon can also take place with other factors such as achievement.

For example, a teacher might be interested in investigating the effects of an instructional procedure that could improve his/her students' performance in arithmetic. He/she decides to test the students before and after instruction. Although he/she does not have use the same test on both occasions, he/she, at least, has to use a pre-test and a post-test that are equivalent in terms of the content areas they represent, their difficulty, etc. But the fact that the students had taken a pre-test that was equivalent to the post-test might have provided them with an opportunity to "practice". Therefore, any improvement in their post-test scores could be due to the fact that they were pre-tested rather than due to the treatment. This type of pre-testing effects is usually referred to as "practice effects."

In summary, it can be said testing effects are the results that the first test has on the performance on the second test.

Instrumentation refers to effects due to unreliable measurement instruments. If one had an unreliable pre-test measurement, any change in the post-test measurement might be due to the instability of the measurement device, rather than the treatment.

Also, when one is using mechanical, electronic, and other sorts of instruments that are in need of calibration or that might get affected by prolonged usage, it is necessary to make sure that any differences between the pre- and post-test are not due to changes in the instruments. In addition, certain research questions might require use of observers rather than tests or instruments. If there is no consistency among the observers or if the same observe is not consistent over a period of time, any difference between pre- and post-test could not be legitimately considered as due to treatment. Therefore, it is necessary to use more than one observer and to make sure that each observer's judgments are highly related with those of other observers (inter-rater reliability).

Statistical Regression is defined as the effects of extreme pre-test scores tending to regress toward the mean on post-testing. For example, if subjects were selected for a study simply because they scored extremely low, and for no other reason, then when they are post-tested, they will tend to score higher, regardless of treatment. Both are examples of extreme cases regressing towards the mean of the population. Therefore, one can get significant differences between pre- and post-test scores just because extremes were selected initially. Regression is a statistical phenomenon that will occur whenever groups are selected on the basis of extreme scores and for no other reason.

Experimental Mortality is the loss of subjects between testing. For example, we can start with 100 subjects on a pre-test with an average achievement of 95 for the sample. If, for some reason, 50 people dropped out before post-testing, the average achievement of the remaining 50 might be 150. Therefore, the difference between pre- and post-test scores would, in all likelihood, <u>not</u> be due to the treatment, but due to the differential loss from pre-test to post-test (mortality).

Selection Bias occurs when subjects are assigned to two or more comparison groups and not all groups are given the treatment. If these groups are initially different before treatment, then any difference between post-test scores may be due to the initial differences rather than treatment. An example of this might be assigning a treatment of individualizing instruction to highly motivated children and traditional instruction to children with little motivation. If the groups were tested for gains at the end of a unit, any difference found might, in fact, be due to initial motivational differences, rather than treatment differences.

The purpose of the "Better" research designs is to eliminate as many of the above type of factors that contaminate research as possible.

These contaminating factors, such as maturation, testing, history, etc., which are threats to internal validity, must be considered when evaluating research. They are useful mind sets to aid an individual in asking the types of questions which will help in identifying good or poor research, and appropriate or inappropriate research conclusions. They are also useful questions to ask before designing or conducting one's own research.

As each design is presented, its strength and weaknesses will be discussed.

Design #1 – The One Shot Case Study (A Pre-experimental Design)

1. X O

In this design, there is some form of treatment to a group followed by an observation of that group after the treatment to ascertain its impact. An example might be a teacher trying a new method of instruction in his/her class and asking the students to indicate how they liked it. Valuable information can be gained, but from a research standpoint, this is not a valid way to assess instruction.

Strengths: Scientifically, this design does not have any strengths because it lacks any form of comparison. Therefore, one can never be sure what the results mean.

Weaknesses: This design is extremely weak in that one really doesn't know what is obtained afterwards since no meaningful comparison has been made. This design is not very often found in the research.

Suggested Statistical Analysis: There is no meaningful way that this design can be statistically analyzed.

Design #2 – The Pre-test, Post-test Case Study (A Pre-experimental Design)

2. O_1 X O_2

This design differs from number one in that an observation is made before the treatment occurs. This allows the researcher to compare the group before and after the treatment to judge the impact of it. For instance, a teacher may pre-test the knowledge of the students related to mathematical concepts. Then, the teacher gives the students a new textbook to read. At the end of the semester, the teacher post-tests the same student with the same test used on the pre-test. The teacher cannot conclude that the new textbook caused any change in achievement.

Strengths: This design is relatively weak. It is stronger than number one, but only because of the one comparison.

Weaknesses: The major weakness is that one can't be sure that the treatment caused the difference between the pre and post measures. On the surface it might appear that the treatment caused a change in the group, but further analysis shows that a number of other factors could be influencing the results.

Suggested Statistical Procedures for Design #2 (O_1 X O_2): Gain scores are most frequently used to analyze Design #2. To do this, the pre-test scores are simply subtracted from the post-test scores to determine gains. A *t*-test or some other statistical test is then used to determine if the gains are significant. However, if one is looking at frequently or nominal data, a Chi-Square, or Sign Test should be used. The most powerful statistical technique to use on this design may be residual gain analysis. This procedure conceptually holds constant gains expected to occur in the post-test, based upon the pre-test scores (slope of the pre-test). The procedure then tests to see if the gains are significantly greater than one would expect from the pre-test regression line.

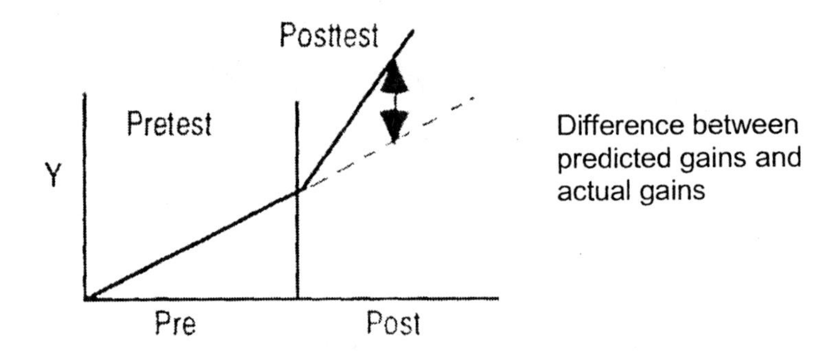

This procedure is statistically more sophisticated and complex, but it also helps to control for the regression effect.

Design #3 – The Simulated Before-After Design (A Pre-experimental Design)

3. $\frac{\quad X \quad O_1}{O_1 \qquad\quad}$

This design uses two groups to enable a comparison of before and after treatment change, but only one group actually gets the treatment. One group is tested before and the second group is tested after receiving the treatment. Interpretation of results is based upon a comparison between the experimental and control groups. The assumption is that the groups are equivalent. One must be careful in interpreting the results of studies using Design #3. Generally, this design is not a very good one and is not used very often.

Design #3 might be used by a teacher who has tried a new approach to instruction and given a post-test. As a control group, the teacher might use another class that happened to get the same test as a pre-test. This design would be normally used when, for some reason, a pre-test was not given to the experimental class.

Strengths: This design does afford some comparisons for the researcher, but the basis for the comparison may be very limited because of uncertainty about equivalence of the groups. Also, this design controls the possibility of a pre-test-treatment interaction since the group pre-tested doesn't get the treatment. Pre-test treatment interaction is an example of testing as a factor affecting the internal validity of a design. In this design there would be no pre-test to affect the experimental group nor in any way sensitize its members to the treatment.

Weaknesses: The major weakness of this design is that one is not sure of the degree to which the two groups are similar (selection bias). This can be a very serious limitation. Other limitations that one should be concerned with are such things are history, maturation, and maybe instrumentation, depending upon the extent to which the instrument may be effected by time.

Suggested statistical procedures for this design are the same as those suggested for Design #2 (see Design #2).

Design #4 – Two Groups – Non-Equivalence (A Pre-experimental Design)

4. $\frac{X\ O_1}{-X\ O_2}$ or $\frac{X\ O_1}{\ \ \ O_2}$

This design involves the use of an experimental and a control group which differs in that one group receives the treatment and one doesn't. It allows a comparison to be made and is employed quite often in educational research. An example might be a teacher using a new method on one class and not using the method on the other class. A comparison is made between the two groups after the treatment is completed.

Strengths: The major strength of this design is that it does allow a comparison between an experimental and a control group. Also, it represents the typical practical situation that exists for many educational researchers—a comparison between tow existing classes. Furthermore, the internal validity of this design is not jeopardized by the confounding effects of history, maturation, testing, regression, and mortality.

Weaknesses: The major weakness is that one can't be sure that the groups are equivalent since there is no attempt to control the composition of the groups. Although this weakness is very severe, it can be lessened by obtaining information that describes the groups on important that describes the groups on equivalence. This is an example of selection bias that was discussed earlier. Another problem one should consider in this type of design is that of *intact groups.* In an intact group, the subjects are not independent of each other. Therefore, one must use the number of groups and not the number of individual subjects in calculating for every design, but especially for designs occurring in natural settings.

Suggested Statistical Procedures: If the data is nominal or frequency, non parametric statistical techniques such as a Chi-Square or Sign Test, would be most appropriate.

If the data is ordinal, interval or ratio, the appropriate statistical techniques would be a *t*-test or an F-test. (It is the personal bias of the authors that parametric procedures are more powerful and are preferable to non-parametric techniques when the data are at least ordinal). This is not consistent with presentations in many traditional texts, but it is based on more than ten years of Monte Carlo studies that generally support this position. These comments pertain to the other designs also.

Design #5 – Experimental-Control, Pre- and Post-test Design (A Quasi-Experimental Design)

5. $\frac{O_1 \quad X \quad O_2}{O_3 \quad -X \quad O_4}$ OR $\frac{O_1 \quad X \quad O_2}{O_3 \qquad O_4}$

Design #5 is a little better than Design #4 because of the inclusion of a pre-test for both groups. However, one still has the problem of nonequivalent groups because the experimenter has not actively designed the groups to be equivalent through randomization or some other means. The pre-test does not provide information about group equivalence that enables a more refined analysis to be undertaken.

This design might be used by an educational researcher interested in two different teaching approaches. The researcher arranges for pre-tests to be given to all classes. Thus, one group, possibly one-half of the classes, receives one method of instruction and the other group, the remaining classes, receives the other method of instruction. At the conclusion of the experiment, a post-test is given to both groups and the results are compared. Also, the researcher can compare the average gains (post-test minus pre-test) for each group or even use the more powerful statistical techniques that quantitatively remove pre-test differences.

Strengths: Although the researcher is not manipulating assignment of subjects to groups, the pre-test gives an indication of how similar the groups are. This design is quite practical in that students may remain in intact classes. This design does eliminate many of the history factors that might influence results as well as maturation. On the whole, the practicality of this design makes its usage widespread despite some remaining questions about internal validity.

Weaknesses: Design #5 is efficient to the degree that the groups are equivalent. To the degree the groups are not equivalent, one cannot assume that the treatment (the independent variable) is causing the difference. For example, if the groups are different, the control group may gain more than the treatment group. In this case, it is possible that the treatment was not effective for the experimental group; but since the groups were initially different, the same treatment may have caused even greater gains in the control group if they had received it. This is referred to by Campbell and Stanley as *interaction*. Therefore, one cannot conclude that the treatment is significantly better or worse, since it may depend on which group received it.

Another possible weakness may be the *regression effect*, if the groups were picked for extreme scores on pre-test. *Mortality* is also a threat to any pre-, post-test design. One should always check to see if there was any differential loss in subjects between pre- and post-tests and whether this loss may account for the difference in gains. The effects of mortality can be assessed by analyzing the number and/or types of subjects that started in both groups and ended in both groups.

Suggested Statistical Procedures: Post-test minus pre-test scores (gain analysis) is generally used. However, analysis of covariance in which the pre-test scores are covaried is the most powerful technique for analyzing this type of data. Details of this procedure can be found in any standard statistical text, such as Popham's *Educational Statistics: Use and Interpretation;* McNemar's *Psychological Statistics;* Edward's *Experimental Design in Psychological Research;* or Ferguson's *Statistical Analysis in Psychology and Education.*

Design #6 – The Longitudinal Time Design (A Quasi-Experimental Design)

6. O_1 O_2 X O_3 O_4

This design is very popular in developmental psychology where researchers are interested in studying growth and maturation, although it can also be used to study learning over a period of time. The distinguishing factor in Design #6 is the number of measurements before and after treatment (there can be more than two). In educational research, this type of design might be employed to investigate the effects of a preschool program. Successive measures throughout elementary school are designed to measure the permanence of such a program. This design can be expanded to include a control group for additional comparisons and increased validity, for example:

5. O_1O_2 X O_3O_4

O_5O_6 X O_7O_8

Strengths: The additional pre-tests(s) allow the researcher to control for the reactivity of the measurements (also called pre-test sensitization). The additional post-tests afford the opportunity of assessing the permanence of gains. Another very important strength is that the multiple observations serve to illustrate the effects of maturation both with and without the experimental treatment. As previously mentioned, the addition of a control group increases the strength of this design considerably.

Weaknesses: A weakness of this design is that the *frequent testing,* especially over a span of years is cumbersome and often subjects are lost for any number of reasons. Another potential threat to the internal validity is *mortality.* In general, this design is mediocre in internal validity, but it is important in longitudinal research. The addition of a control group makes it a good design with a fair amount of internal validity.

Suggested Statistical Procedures: One method used frequently with this type of design is simply graphing on the X-axis the observations (independent variable) and on the Y-axis the criteria measures (the dependent variable).

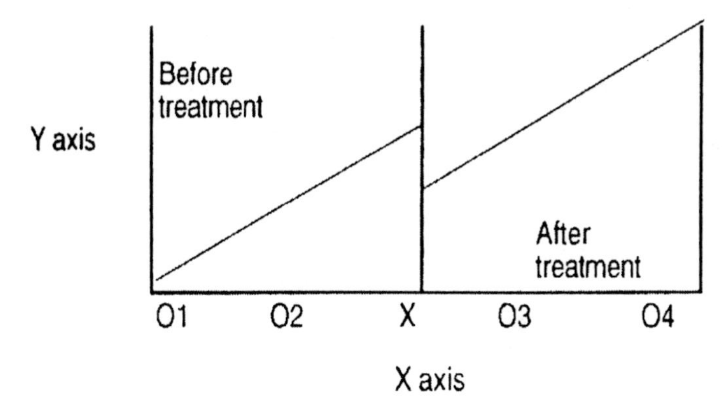

One should avoid the tendency to average all the pre-test score and test them against the post-test scores. This can be very misleading.

The most appropriate way to analyze this design requires a fair amount of statistical sophistication. The researcher looks at the intercept and slope of the line before treatment (X) and compares it to the intercept and slope of the line after treatment (X). This type of analysis is specifically appropriate for behavior modification designs. (See Kelly, Newman, and McNeil, 1973)

Since there are underlying statistical difficulties in analyzing such data, it is a good idea to consult a statistician before the analysis.

Design #7 – Experimental and Control Groups Randomized Subjects (A True Experimental Design)

7. R X O_1 R X O_1

OR

R -X O_2 R X O_2

This design represents the classically designed experiment. The groups are equated by randomization and one group is randomly (R) designated as experimental and the other as control. The experimental group receives a treatment, the control group doesn't and the two groups and compared after the treatment to assess its impact. An example would be a school investigating a non-graded program on an experimental basis. One-half of the first graders are assigned by random means to experimental classes. The remaining students are assigned to control classes. At the end of the school year, comparisons are made between the two types of instruction. In this example, the researcher would probably measure more than one variable on the post-test. Achievement, attitude toward school and other important variables could be compared.

Strengths: Design #7 is considered to be an outstanding design from a scientific point of view. Its main strength lies in the fact that the control group enables a legitimate comparison to be made between groups. This is the first design observed so far that incorporates the equivalent control group into the design. Related to the idea of the *equivalent control group* is the technique of randomization. *Randomization* is the procedure that makes this design effective. When subjects are randomly assigned to treatments, one is allowed to assume that the groups are equivalent on all possible variables and differences between groups can be assumed to be the result of chance alone. This design provides adequate controls for all threats to internal validity.

Weaknesses: The major weakness of Design #7 is in the area of external validity. The problem is that most classes in schools are not constituted randomly. Therefore, the application of research results based upon Design #7 is only as valid as the process used in obtaining groups that allows one to use randomization procedures. When students are grouped homogeneously into tracts, to some the degree they are different from students that are grouped with no bias whatsoever. This problem can be solved by making classrooms the unit of observation instead of individual students. This solution necessitates an extremely large study since sample size would be determined by the number of classes in the study.

Another weakness of this design is that it is usually quite difficult to randomly assign subjects in many applied settings such as schools. The weakness is usually a matter of practicality in the normal routine of school operations.

Extensions: This design can be extended to include studies with more than two groups in which randomization is used. As more groups are added to a study, one finds that the notion of a control group (no treatment) may become inapplicable. It may become convenient to think of these groups as comparison groups, each getting different treatments, instead of the classical experimental-control group design. The logic of the investigation is still the same: groups are treated exactly the same on all except one variable whose effect is observed on some criterion. This design is the simplest form of the post-test only control group design. Further examples would be:

$R\ X_1\ \ O_1$ OR R $X_1\ O_1$, $X_2\ O_2$, $X_3\ O_3$ OR R $X_1\ O_1$, $X_2 O_2$, $X_3\ O_3$, $X_4\ O_4$ OR $R\ X_1\ O_1$, $R\ X_2\ O_2$, $R\ X_3\ O_3$ etc.

$R\ X_2\ \ O_2$

Suggested Statistical Procedure: If the criterion variable is a frequency (nominal data), then a Chi-Square, Sign Test, or similar procedure is suggested for statistical analysis. If the data is ordinal or interval in nature and there are only two groups, then a t-test is appropriate. If there are more than two groups, then a *t*-test is appropriate. If there are more than two groups, then an F-test would be most appropriate. However, depending upon the number of groups and the research question, analysis of covariance or blocking procedures may be used to increase the power of the statistical analysis.

Design #8 – Experimental and Control Groups Matched Subjects (A True Experimental Design)

8. M_r $X\ O_1$, $-X\ O_2$ OR $M_r\ X\ O_1$, $M_r\ X\ O_2$

Design #8 is quite similar to #7 except in the method used to assign subjects to groups. In Design #8, subjects are assigned to groups by matching them first on some attribute or attributes. In addition, it is important to note that, as each pair of matched subjects is selected, the assignment of members of the pair of groups is done randomly. If a researcher wanted to study the effects of lectures vs. independent study on achievement in biology, for example, subjects would probably be matched on previous biology achievement. If subjects were matched on previous biology achievement, for example, the researcher would then randomly assign the members of each matched pair in the sample.

However, one of the major problems with matching is the difficulty in obtaining subjects. Therefore, one may lose some power because many unmatched subjects may have to be eliminated from the subject pool.

Randomization is generally regarded as the best technique for equating groups because of its greater precision and relative ease in implementing.

Suggested Statistical Procedures: The statistical procedures are the same as those suggested for Design #7. However, if analysis of covariance is used, one can covary (hold statistically constant) the variable that was used for matching, allowing more subjects to be used. Therefore, more power is obtained without having to match.

Design #9 – Before and After, With Control Groups (A True Experimental Design)

9.								
	O_1	X	O_2		R	O_1	X	O_2
R				OR				
	O_3	-X	O_4		R	O_3		O_4

Design #9 includes the use of a pre-test with groups that have been equated by means of randomization. The pre-test results will demonstrate to the researcher the effectiveness of randomization as a means of equating groups. In addition, the pre- and post-measures allow the researcher to assess the impact of treatment using gain scores. Of course, the use of equated groups in this design allows comparisons to be made between experimental and control groups.

An example of Design #9 would be research on the effectiveness of a new mode of instruction, for example, programmed instruction vs. traditional, on end of year achievement in arithmetic. Students are randomly assigned to two groups and given a pre-test at the beginning of the school year. The experimental group of classes uses programmed instruction in arithmetic and the control group does not—other conditions are hopefully the same for each group. At the end of the school year, all students are tested in achievement again. Usually, the same test or a parallel form of it is used. The researcher then compares the average gains during the year to see which group has gained the most.

Strengths: This design is very strong in internal validity. The use of both randomization and the pre-test provides sound evidence of the history and maturation. Also, the ability to analyze gain scores is welcomed in some research situations, especially when assessing achievement over the period of a school year.

Weaknesses: The major weakness is the practical difficulty of using randomization in many applied settings. Sometimes, it is just not practical in a school situation. Also, the use of a pre-test may cause some disruption of normal activities and might predispose subjects to react unusually to the treatment. For example, a pre-test of racial attitudes might affect the subjects' sensitivity to an experiment investigating the impact of encounter groups on racial attitudes (threat to external validity). Likewise, an achievement test at the beginning of a school year might affect students' and teachers' motivation toward an arithmetic course. This phenomenon is called pre-test sensitization. In comparison to the other designs, however, the weaknesses of design #9 in terms of internal validity are minor.

Suggested Statistical Procedures: The procedures suggested for this design are the same as those suggested for Design #5.

Design #10 – The Solomon (A True Experimental Design)

10. (Group I)	R	O_1	X	O_2
(Group II)	R	O_3		O_4
(Group III)	R		X	O_5
(Group IV)	R			O_6

Design #10 is the most elegant of the designs in this chapter in terms of the internal validity that is insured by it s use. In essence, it is really a combination of Designs #7 and #9. The Solomon, named after its originator, allows the researcher to make a number of comparisons. Again, using randomization insures that the comparisons rest on sound ground. History, maturation, and pre-test sensitization are all controlled in this design.

An example of its use would be a study of the effect of watching a movie on drug abuse upon attitudes toward drugs. Groups I and II and administered a pre-treatment questionnaire which assesses attitudes toward drugs.

Groups III and IV are not given the pre-test so that any possible pre-test sensitization will be evident in comparison between Groups I and II and Groups III and IV. At the end of the movie on drug abuse or the control group's movie on recreation, the drug questionnaire is administered again but this time to all the groups. Differences between Groups I and III compared with Groups II and IV show the impact of the film on drug abuse attitudes. Remember that all four groups were equated before the experiment by randomization.

Strengths: This is by far the strongest design for the control of internal validity. Group equivalence through randomization and the four different groups allow comparisons to be made which answer all the internal validity questions.

Weaknesses: The major weakness with Design #10 is its practicality. The Solomon Design is difficult to set up in many situations. Randomization, four different groups, and some pre-testing, are factors that make Design #10 difficult to implement. The difficulty of implementation detracts from the usefulness of this design. An additional weakness is that, because this design has four groups, it requires twice as many subjects as Designs #7 or #9, which only have two groups. Since this design requires procedures that are unlike the routine of the average school, generalization to the average school is limited.

Suggested Statistical Procedures: One method of statistical analysis is to put this into a 2 x 2 analysis of variance design. See Illustration #9 for an example of this design.

Illustration 9: Example of a 2x2 Analysis of Variance Design

		(Treatment)	
		yes	no
(Pretest)	yes	Group I	Group II
	no	Group III	Group IV

This design gives information about the effects of the treatments and the effects of pre-testing as well as the effects of the treatment interacting with the pre-test (i.e. A – main effects, B – main effects, AB – interaction).

Chapter VIII

Type VI Error: Inconsistency Between the Statistical Procedure and the Research Question*

Isadore Newman, Robert Deitchman, Joel Burkholder, Raymond Sanders, The University of Akron; Leroy Ervin, Jr., Oberlin College

While a great deal of money and energy is currently being directed toward research, there also seems to be a general lack of acceptance of the relevance of research findings. This skepticism is somewhat justifiable, and we will attempt to discuss what we feel are the major causes of this state of affairs. These issues may explain to some extent why less and less monies are being made available for certain types of research.

One reason people ignore research findings is that the statistical models used have frequently been unrelated or tangentially related to the research question of interest. There are a variety of reasons for this, some of which will be discussed in this paper.

1. The courses that teach research methods generally emphasize data analysis rather than practicing appropriate methods and procedures for asking and developing research questions. These courses do not adequately develop the skills of evaluating the research question and the statistical models that are most capable of reflecting the research question.

Quite often students coming out of these courses tend to select a familiar, "canned" standard statistical design, or package (cookbook approach) such as a 2 x 3, or one 2 x 2 x 3, because they have not been taught to develop their own models to reflect the research question. Therefore, they use these standard models which dictate the question being investigated. Sometimes a researcher is aware that these models do not completely represent the true research question. When this happens, he or she may then make inferential jumps from the data. These influences may well be inappropriate. In some cases, the researcher is unaware that the models are not really reflective of the research questions; sometimes the unsophisticated researcher allows the statistical model to totally dictate the research question. Under these conditions, we find research that is technically correct but is not relevant because it is not related in a pragmatic way to a specific problem.

2. People often misinterpret or misapply research findings because they confuse statistical analysis with experimental design. Traditional analysis of variance, in which the design and statistical procedures are totally related, has fostered this confusion.

One sometimes finds even the most respected of colleagues making statements such as:

a. We can only infer causation when the research is analyzed through traditional analysis of variance.
b. If we use regression or correlation, we cannot infer causation.

Remember, causal relationship can only be inferred through experimental design. No statistical technique allows us to infer or assume causal relationships.

In an issue of the *Harvard Review*, Luecke and McGinn (1975) wrote a paper entitled, "Regression Analyses And Educational Production Functions: Can They Be Trusted?" Their conclusion was that we cannot appropriately infer causation from regression techniques. This is an obvious statement since causation cannot be inferred from any statistical technique, whether it be regression, analysis of variance, t-test, etc. However, their conclusion has been inappropriately generalized beyond this. In some quarters it has been accepted to the extent that some believe causation cannot be inferred anytime regression is used regardless of the research design. This stems from the earlier problem of confusing statistics with design.

Regression is a statistical procedure and should be used in relationship with some design in the Campbell and Stanley (1969) tradition. A researcher should begin by asking a research question that reflects an area of interest. The study should then be designed so that it can answer that research question. The research design should include how data will be collected, how subjects will be selected, how treatments will be administered, etc. (see Fig. 1). To the extent that this design has internal validity, one can infer causal relationships between the independent variables. If this design is internally valid, no matter what statistical technique is used to test it, such as regression, correlation, etc., one can legitimately assume causal relationships.

If the research design is ex post facto, i.e. where the independent variable is not under the control of the researchers, no matter what technique is used, one could not infer causation. All the studies on desegregation and busing, such as the Coleman Report (1966), are of this nature. It is not technically legitimate to infer causation when the design is ex post facto. Even though a variety of statistical techniques such as path analysis as developed by Blalock (1970) and component analysis developed by Mood (1971) have attempted to get at causal relationships of ex post facto data through the manipulation of regression techniques, one still cannot technically infer causation. Causation cannot be inferred from an ex post facto design even if you analyze data with ANOVA. These techniques tend to eliminate some rival hypotheses, making the study somewhat stronger. However, they cannot eliminate all of the rival hypotheses so the study cannot be totally internally valid and causation cannot be inferred. This is not to say that these studies are not very valuable. This simply says one cannot infer causation between the independent and dependent variable.

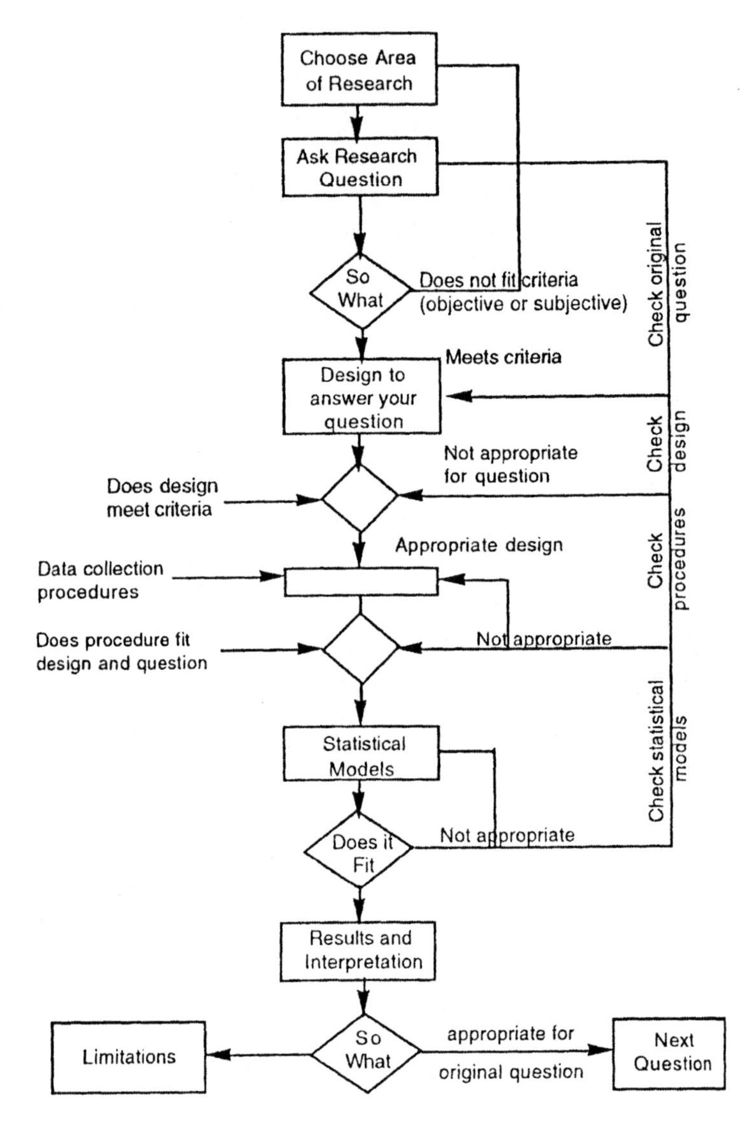
Choose Area of Research
Ask Research Question
So What
Does not fit criteria (objective or subjective)
Check original question
Meets criteria
Design to answer your question
Not appropriate for question
Check design
Does design meet criteria
Appropriate design
Data collection procedures
Check procedures
Does procedure fit design and question
Not appropriate
Statistical Models
Check statistical models
Does it Fit
Not appropriate
Results and Interpretation
Limitations
So What
appropriate for original question
Next Question

Figure 1

A necessary but not sufficient prerequisite for an experimental design is that at least one independent variable is active (under the control of the researcher). If there are no active variables in a research study, the study by definition would be ex post facto, even if the analyses is ANOVA. For instance, assume one study with a 2 x 2 factorial design looked at Black students in integrated schools versus Blacks who were not, and sex. If the subjects were not in the researcher's control, this study would then be an ex post facto design not a true experimental design since random assignment could not be employed.

	Male	Female
Integrated		
Non-integrated		

In effect, this is a straight correlational study even though one is getting A-main effects, B-main effects, and AB-interaction. It is not a true experimental design, and causation cannot be inferred. The statistical question being asked in such a design is whether there is a significant relationship between these independent variables and the criterion. Any inference about causal relationships would be inappropriate.

3. When the research question of interest is one of trends or functional relationships, one often finds the use of inappropriate statistical models that cannot accurately reflect the research questions (Newman, 1974).

When researching developmental questions, one is often more interested in functional relationships than mean differences. There is generally a continuous variable that is of interest, such as time, age, population sizes, IQ, sex. When traditional analysis of variance is employed, continuous variables are forced into categorizations.

This causes the researcher to lose degrees of freedom, and there is a potential loss of information. This loss is contingent upon how representative the categories are of the inflections in the naturally occurring continuous variable.

Since continuous variables are frequently artificially categorized, the analysis produced by such a procedure may not really reflect the researcher's question or interest. The most efficient method for writing statistical models that reflect trend or curve fitting questions is the general case of the least squares solution, linear model (Multiple Linear Regression Procedures, Newman (1974), Draper & Smith (1966), Kelly, Newman, and McNeil (1973). This procedure allows one to write linear models which specifically reflect the research question.

Linear Regression is an excellent statistical tool for looking at a population trend or comparing multiple trends over time (Newman 1974). The following examples are taken from a study comparing performance trends of under-prepared students receiving special treatment with the performance of regularly admitted students requiring special assistance over four years (Ervin, 1975). Although the example provided is concerned with an educational issue at a single institution, regression is not limited to such situations. In fact, it has more flexibility than of any other single statistical tool and can be used with large populations incorporating a sizable number of independent variables in a single model.

In Figure 2, a graph is presented that reflects the researcher's interest in learning if there are significant differences in trends (in this case slope differences) between Black subjects who received the Developmental Program (X_2) and Black students who did not receive the program (X_1), as it relates to their cumulative GPA.

This research question was then related as a specific hypothesis.

Hypothesis I : Are there significant differences in slopes for X_1 vs. X_2 in predicting cumulative GPA for Black students at Oberlin?

The models needed to test this hypothesis are as follows:

Model 1: $X_{14} = a_oU + a_1X_1 + a_2X_2 + a_3X_3 + a_4X_4 + E_1$

Model 2: $X_{14} = a_oU + a_1X_1 + a_2X_2 + a_5X_5 + E_2$

Where: X_{14} = cumulative GPA

X_1 = 1 if student had program, 0 otherwise

X_2 = 1 if student did not have program, 0 otherwise

X_3 = number of the semester for the subjects who had the program, 0 otherwise

X_4 = number of semesters for the subjects who did not have the program, 0 otherwise

X_5 = $X_3 + X_4$ = number of semesters for all subjects

U = unit vector, 1 if subject is in the sample, 0 otherwise

$a_o - a_5$ = partial regression weights

E = error Y-Y

If Model 1 is found to be significantly different from Model 2, this would indicate that there is a significantly different functional relationship (at some specific level) between Black students who took the program and Black students who did not take the program in terms of their cumulative GPA.

There are an infinite number of other questions that could be asked, such as: Are there significantly different 2nd degree or 3rd functional relationships (curvilinear relationships) and are the means of the groups significantly different over all semesters or over specific semesters? etc. Then, regression models could be written that specifically reflect the questions of interest.

4. The problem of unequal Ns is really an intriguing one. It is of importance because it is a common problem for applied researchers; and, depending upon the models employed, it can produce widely different answers.

There are a variety of solutions to the unequal N's problem that can be divided into two major categories—approximate and exact. Examples of approximate solutions are: randomly eliminating data and running the analysis on just group means, therefore, decreasing the number and power. A researcher using any of these solutions is generally aware of the limitations and problems.

What may be more misleading are the exact solutions which are all technically correct but which, like the mean, median, and mode, are answering different questions. The three exact solutions are the hierarchical model, the unadjusted main effects method, and the fitting constants method.

1. *Solution I* (Cohen, 1968; Williams, 1974). This requires a priori knowledge of which variables in a solution are more important or of greatest interest. With this approach, each effect is only adjusted for the effects that are of greatest interest.

2. *Solution II* (Williams, 1974). This solution requires interaction effects would be adjusted for the variance accounted for by the main effects.

3. *Solution III*. This method as a solution for disproportionality has been suggested and described most frequently (Winer, 1971; Scheffe, 1959; and others). In this procedure, each main effect is adjusted for each other main effect, and the interactions are adjusted for all main effects. This procedure is similar to the analysis of covariance procedure.

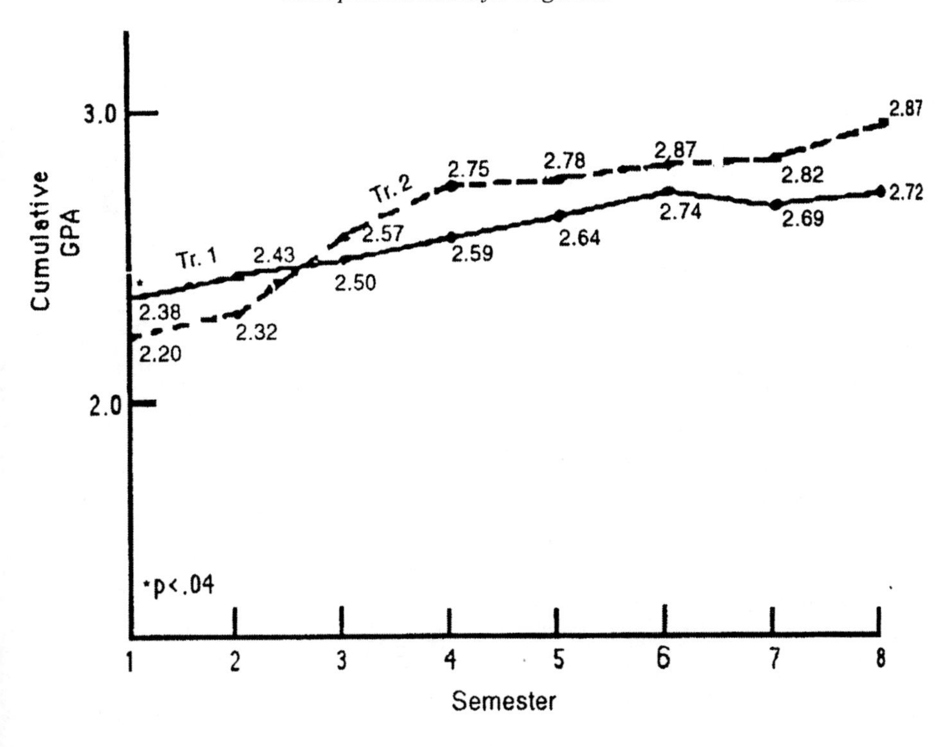

Figure 2

By examining Table 1, one can see that the hierarchical method is most sensitive to detecting questions of major interest. It accounts for the total variance of the variance of the main effect, even though it overlaps with the variance of each main effect minus the variance of the other main effects. The third method of fitting constants may be most conservative of the three. It determines the unique variance accounted for by each main effect and the interaction. As one can see from Table 1, these three methods, which are all statistically accurate, are answering three different questions.

It is important to determine which of these methods are reflecting the question that we are interested in answering. One can only do this by being sensitive to one's research question and by being aware of the different statistical techniques that are more appropriate than others.

4. Still another area of concern for applied researchers is the relationship between statistical significance and N size. When the Ns are very large, any difference is likely to be statistically significant at extremely conservative alpha levels. It is the authors' position that it is to the advantage of the researchers and the implementers of research to always look at the R^2 (proportion of variance accounted for). This is a very important index to estimate the pragmatic importance of the study. This becomes especially important when the Ns are large and one finds significance at $p < .001$. It is possible that the R^2 is accounting for 1% of the variability and 99% is unaccounted. It is important that the researcher be aware of the magnitude of effect as well as the statistical significance. Without this awareness, one increases the likelihood of inappropriately estimating the usefulness of the results.

5. Another error that sometimes occurs due to inappropriate statistical procedures is the loss of power (increase in Type II error, failure to reject a false null hypothesis). This may be due to not using a directional hypothesis when one should, especially when the N is small.

A relatively frequent error found in the literature is the statement that, "Significance was found, but in the wrong direction." This is incorrect because if it was in the wrong direction, one would have to fail to reject the null hypothesis; and therefore, conclude the findings were nonsignificant.

A possibly more dangerous error, because of its subtlety, is when a researcher makes a nondirectional test to "play it safe" and because most statistical textbooks suggest this procedure. He or she then might find statistical significance in the opposite direction from which the question was initially generated. (In other words, the researcher should have used a directional text, but did not.) Some researchers would then state that they have found statistical significance, but it is a *conceptual error*. In this case, statistical significance is likely to mean that there was something wrong with the initial logic or theory from which the hypothesis was derived, and a rethinking of the initial logic and procedure is called for. Therefore, this aspect must be carefully looked at before one can determine appropriate estimates of the findings' applicability. (See McNeil et al. (1975) for an excellent discussion on the importance of directionality in hypotheses testing.)

Suggested Approach for Conducting Research

It has been suggested that the sequential steps for conducting research, which are presented in Figure 1, be adopted as a guide (Deitchman, Newman, Burkholder, and Sander (1975).

At best, one should always generalize with caution. The magnitude of the relationship that one finds in a sample is quite often an overestimate of the magnitude of the effect in the population. Therefore, one should try to replicate whenever possible.

If one does not wish to or cannot replicate, the next best procedure of estimating the effect size in the population from a sample is by using shrinkage estimates. Shrinkage formulas are most appropriate as unbiased estimates of R^2 when the predictor variables are not solely selected on the basis of their initial correlation with the dependent variable.

One of the best estimates of shrinkage is a cross validation procedure. This procedure is described in detail in McNeil, et al. (1975). Three easily computed mathematical estimates of shrinkage are discussed and evaluated by Newman (1973) and Klein and Newman (1974).

Table 1

CONCEPTUAL PRESENTATION OF THE THREE METHODS FOR ADJUSTING UNEQUAL N'S

SOLUTION I	Total proportion of variance accounted for by the effect of major interest. Example: (n^2 for A = main effect)	Proportion of variance accounted for by the effect of the second most important variables holding constant the first most important variable. Example: (n^2 for B holding A constant)	Proportion of variance accounted for by the last important variable holding the first and second most important variables constant. Example: (n^2 for ns holding A and B constant)
SOLUTION II	Proportion of variance for one main effect holding the other(s) main effect(s) constant. Example: (n^2 for A-main effect holding a main effect constant)	Proportion of variance accounted for by the second main effect holding the other(s) main effect(s). Example: (n^2 for B-main effect holding constant A-main effect(s)	Proportion of variance accounted for by the interaction while holding constant all the main effects. Example: (n^2 for AB holding constant A & B main effects)
SOLUTION III	Unique variance accounted for by the first main effect holding constant the other main effect(s) and interaction(s). Example: (n^2 for A-holding constant B & AB)	Unique variance accounted for by the second main effect while holding constant the other main effects and interaction(s). Example: (n^2 for B-effect holding constant constant A & AB)	Unique variance accounted for by the interaction holding constant all the main effects. Example: (n^2 AB holding constant A & B main effects)

In addition, anyone who is conducting or reading research should keep in mind the important difference between statistical methods and research design. As depicted in Figure 1, the first important step is to ask a relevant research question and then to decide on procedures and methods (design) which will lead to an answer for that question. This design should then be evaluated for limitations in terms of its internal and external validity. Once one establishes the basic research question and a design that is acceptable, specific hypotheses can be derived which can be statistically tested (keeping in mind that every test of significance is a test of specific hypothesis.) Linear models (regression models) can then be developed to reflect each specific hypothesis and to test each for significance. In this manner, when one gets statistical significance from a test, one knows specifically the hypothesis being tested. He or she also knows how accurately the results can be interpreted in terms of causation and its general ability, based upon the internal and external validity of the research design.

In Retrospect

We have attempted to convince trainers of researchers and the researchers themselves to beware of the Type VI Error. We believe that the examples will help the researcher to become sensitive to these types of problems. Bad research not only drives out good money, but also creates a large credibility gap where there should be none. From the authors' perspective, applied and theoretical research really only differ in terms of face validity that implies an attempt to answer the "So What" question of the research. It is to some extent an artificial dichotomy that applied researchers are dealing with the "So What" and the theoretical researchers are not. In reality, this is not true. Asking the "So What" question may be one of the discriminating features between good and bad research, regardless of orientation.

The authors of this paper feel that good research can benefit everyone, but poor research may harm everyone. We also believe that the suggestions presented in this paper, if followed, are likely to facilitate good research.

Chapter Summary

A Type VI Error was defined as the inconsistency between the researcher's question of interest and the statistical procedures employed to analyze the data. The reasons for these commonplace errors vary. They may be the result of courses which fail to emphasize logical analysis as a prerequisite to the statistical analysis, confusion caused by teaching traditional analysis of variance without differentiating between the statistic and the design, the misconception that causation can only be inferred from traditional analysis of variance, and forgetting that causation is a design question, not a statistical one.

A major problem discussed was disproportionality (unequal Ns). The three exact solutions, hierarchical models, unadjusted main effects, and fitting constants, were discussed in terms of conceptual questions and answers. Stress was placed on understanding that each solution is answering different questions. Therefore, one should not be concerned with which model is more statistically accurate, but with which one is more reflective of the researcher's question.

In dealing with a research question, one must consider practical as well as statistical significance. This means that N size, and effect size as well as probability level and power, must be considered.

Finally, a procedure is suggested for conducting research along with a flowchart. We feel that following the procedures outlined in the flow chart will help researchers avoid some of the pitfalls that lead to Type VI errors.

---

*Paper presented to the World Population Society, Western Regional Meeting; January 24, 1976, aboard the Queen Mary, Long Beach, California.

We would like to gratefully acknowledge the assistance of Dr. Ralph Blackwood, Dr. Keith McNeil, and Dr. Charles Dye for their editorial comments.

be somewhat subjective, that
between the system used in the
interest. It is important to note th
estimate can be calculated befo
is implemented. If it is calculated
study, it could assist in the re-
actually implemented. If co
implemented, it will have impli
decision makers.

Statistically Significant
Study

One value we believe shoul
is the likelihood that the study's
system. Such a value should
significance tests but rather sho
We agree with Robinson and
position that the probability
statistical test is an important p
quantitative study. Robinson an
first indicate whether the c
improbable one (e.g., is the di
be expected by chance?)" (p.

It is important, however, r
respect to the likelihood of th
2000). This point was addressed
"some believe [incorrectly] tha
that at least 95% of replicatio
(p.102). Posavac does take the
null hypothesis should increase
replications of the research wo

Using p Values to Estimate Statistically Significant Exact Probabilities

Greenwald, Gonzalez, Harris, and Guthrie (1996) presented an analytic method by which a *p* value can be converted into a probability estimate that an exact replication of the research would produce a statistically significant result. Posavac (2002), who elaborated on the method proposed by Greenwald, et al., noted that an "*exact replication* means that the initial experiment is repeated using the same independent and dependent variables with the same number of participants selected in the same way from the same population" (p.102). In this type of replication the difference between the replication and the original study is due to random variation. We believe that this is one type of replication that should be addressed by researchers.

Green et al. (1996) and Posavac (2002) suggest that the probability of a statistically significant exact replication (SSER) can be estimated from the probability of the statistical test. As a means of demonstrating how a researcher could convert a *p* value from a statistical test to an estimate of the SSER probability, we present a brief discuss of the procedure. It is beyond the scope of this paper to present the rationale on which this procedure is based. We encourage interested readers to review the works published by Greenwald et al. and Posavac for a more in-depth discussion of this concept.

An Illustration

To illustrate the calculation of the SSER probability value, assume researchers are testing the difference between sample means of two independent groups consisting of 20 subjects each. Further assume that the *t* value produced by the difference between the two means recorded for their study was 2.150. Since this observed *t* value (t_{obs}) is greater than the two-tailed critical *t* value (t_{crit}) of 2.024 for an alpha level of .05, the researchers would declare the difference between the two group means to be statistically significant. The question we believe is important for these researchers to address is: What is the chance that the difference between the two group means recorded for an exact replication of the study would be declared statistically significant?

Calculation of the SSER probability value

As noted by Posavac (2002), the probability of obtaining a SSER can be obtained by executing three steps. First, the replication *t* value (t_{rep}) is calculated by subtracting the critical *t* value used in the initial study from the study's observed *t* value. Thus the t_{rep} value is calculated as follows for our hypothetical example:

$$t_{rep} = t_{obs} - t_{crit}$$

$$t_{rep} = 2.150 - 2.025$$

$$t_{rep} = .125$$

Second, the researchers would obtain the one-tailed probability for this t_{rep} value of .125 with 38 degrees of freedom. With respect to the procedure used in this step, Posavac (2002) stated that "a one-tailed test is used because one would want a replication to produce means of the same relative magnitudes as found in the first study (p.108)." The one-tailed probability for the t_{rep} value of .125 with 38 degrees of freedom is .45.

Third, the researchers subtract the .45 probability value from 1.00, which produces a value of .55. this value indicates that the chance that an exact replication will be statistically significant is .55.

Points to note regarding the SSER probability value

Three points should be noted regarding this SSER probability value of .55. First, the SSER probability value is a function of the *p* value. However, practitioners need to be careful not to directly interpret the *p* value as a replicability value. Second, Greenwald et al. (1996) and Posavac (2002) recommended that SSER probability values should be considered upper limits. The reason for this recommendation is based on the fact that "even in a careful replication the partcipants would be a different sample from the population, the calendar date would be different, the weather would be different and so forth" (Posavac, p.111). Third, researchers may be surprised that for a study, such as the one used in our example, which had 38 degrees of freedom and an observed t value of 2.150 ($p = .038$), the chance that an exact replication will be statistically significant (SSER probability level = .55) is only slighlty above the 50-50 level. In fact an observed t value for this hypothetical study would need to be 2.874, which produces a p value of .01, in order for the SSER probability level to reach the .80 level.

To further emphasize this third point, a review of values produced by Posavac (2002) reveals that when degrees of freedom value is at least eight and the p value is .05, the SSER probability value will be .50. That is, there is a 50-50 chance of replicating significant findings. If the degrees of freedom value is at least eight and the p value is .01 for a two-tailed test, the SSER probability value will not be less than .73 or greater than .84. And if the degrees of freedom value is at least eight and the p value is .005, the SSER probability value will not be less than .80 and not greater than .92. (It is interesting to note that these replicability values are less for corresponding p values for one-tailed tests.) Thus researchers need to be careful not to assume that statistically significant findings automatically mean that the chance of obtaining statistically significant exact replications for the study will be high. For this reason we believe that researchers should report the SSER probability value along with the probability of the observed t test.

Replication in a Different System

We believe that a second type of replication of findings is important for researchers to address. That is, the type of replication that deals with the question: Would the study's findings replicate in a system different from the one used in the initial study? It should be noted that we consider this type of replication of findings important even if an individual is interested in the same system in which the study was conducted, assuming the system is a dynamic one. That is, the system experiences considerable change with respect to the variables that may influence the replicability of the findings. Since most people attempt to relate research findings to systems that are different from their own or, at least, relate findings to systems that are similar but dynamic, we believe obtaining a likelihood estimate for this type of replication would be most valuable for them. The remaining portion of this section of the article presents our *preliminary* attempt to develop a procedure for calculating such a likelihood estimate.

Estimate Procedure

The procedure we are proposing for the estimate of the likelihood of replication of findings for a system different from the one in which the study was conducted can best be presented through an example modeled on a study conducted by Benson, Aronson, Desmeet, Shaheen, and Showalter (2002), which presented an evaluation of a multiage classroom educational program. In our example we have children in grades 1-3 who were grouped in the same classroom and their teacher stayed with them for three years. The evaluation indicated that the teachers volunteered for the project and were enthusiastic about the concept of multiage education. The project was supported fully by the principals and was enthusiastically supported by the parents. Achievement scores indicated moderate success of the project as compared to national norms and comparison students in the same school.

Internal validity issues are apparent, since parents voluntarily allowed their children into the project (see Campbell and Stanley, 1963, for a discussion of internal validity issues). Enthusiastic teachers might generate better results, no matter what the curriculum. In addition, a supportive principal might be partly (or entirely) responsible for the achievement results. Other internal validity concerns could also be raised.

External validity issues are also of concern with this study. Would the same effects be observed with less enthusiastic teachers? Will the same effects occur after the novelty of the multiage grouping wears off? Other external validity concerns could be raised (see Campbell and Stanley, 1963, for a discussion of external validity issues). The concept of replicability, though, is different from internal and external validity. It is based on the realization that that system is likely to be dynamic. We believe that the likelihood of replicating a study's findings in a different system or even the same dynamic system is crucial to estimate.

Important variables

The first step in the estimation process is to identify key variables that influenced the findings but may be different in the new system. Let us assume that for our multiage project example four such variables were identified:

1. Twelve volunteer teachers were used.
2. The study involved supportive principals.
3. The study used 240 volunteer (supportive) parents.
4. A total of 5 days of in-service training was given to the teachers on the multiage project.

As an illustration of how these variables could influence the replicability of the findings of the original study, consider the principals variable. If a principal leaves, the project will, in all likelihood, be supported less by the new principal. The new principal may even kill the project, not because the project is ineffective, not because the concept of multiage education is bad, but because the crucial component of the system (the principal) does not believe in or want the project.

The likelihood of each crucial component changing should be taken into account when the project is envisioned. If a particular component is likely to change, then the project should be devised so it is immune to that change--in the case of principal change--the project should be made "principal proof."

Once the variables are identified, the second step is to estimate the proportion of the R^2 value accounted for by each variable, the probability of that variable changing, and the probability of the changed variable being negatively influential on the original findings. Table 1 contains such hypothetical values of these estimates for our example.

Table 1:
Proportion of the R^2 Accounted for in the Dependent Variable and the Probability Values for Each Variable

Variable	Proportion of R^2	Estimate of the Probability of Change	Estimate of the Probability of Negative Change
Teacher	.50	.30	.03
Principals	.20	.70	.60
Parents	.10	.33	.02
Staff Development	.20	.02	.02

The proportion of the R^2 value accounted for by each component is determined. This could be accomplished with GLM if enough implementation sites were available (similar to meta analysis), or conceptualized either before the study started or afterwards. In the example, here we provide "educational guesses." For instance, it is likely that some teachers will leave the project. Some may become disillusioned with the project or with education in general. Others may find a more lucrative job in another district or another profession. Nevertheless, other enthusiastic teachers are likely available, so the systemic effect on the project of teacher change would be minimal.

On-the-other hand, the likelihood of a principal leaving the system is high (estimated to be .70 in a three-year period) and the likelihood of the replacement being equally enthusiastic is low

(.40). Indeed, most replacement principals may gut the project, leading to absolutely no replicability from the component of the principal. Therefore, because of the high probability of principal change, and high probability of a different (lower) level of support, the overall replicability is lowered.

Parent turnover will be at least 33% every year, with third graders moving to fourth grade. But we suspect that the parents of the incoming first graders will be just as enthusiastic (maybe even more so if the project is a success). Thus, the high turnover rate (large system change) of parents will have little effect on replicability--the project is "parent proof." If the staff development is "packaged" then it could easily stay the same from year to year. This part of the system would likely be stable.

Actual probabilities may be quite difficult to determine. To deal with this problem, one might rate the stability of each component on a 1 to 5 scale, with 5 being the most stable. Such estimates and the calculation of the reliability value for the multiage example are listed in Table 2. It should be noted that a replicability value calculated in this manner would produce higher values the more stable key variables are from the system used in the initial study and the system of interest especially for the variables that account for the higher proportion of the R^2 value.

Table 2:
Calculation of the Replicability Value

Variable	Proportion of R^2	Stability	(Proportion of R^2) * (Stability)
Teacher	.50	4	.50 * 4 = 2.00
Principals	.20	1	.20 * 1 = 0.20
Parents	.10	5	.10 * 5 = 0.50
Staff Development	.20	5	.20 * 5 = 1.00
		Replicability = 3.70/5 = .74	

Estimating Replicability Before Implementing

Minimizing the effects of change in teachers could be accomplished by each project teacher identifying a non project teacher who would like to be in the project and then keeping that teacher informed about multiage grouping during the year. This "information partnership" actually becomes a new component of the project (or at least modifies the teacher component). Curricula that purport to be "teacher proof," such as highly prescriptive direct instructional methods, are another example of minimizing the effects of teacher change.

If the replicability index is low, and the researcher cannot identify changes or strategies that would make it higher, then the project should not be implemented. The time of teachers, principals, parents, staff developers, and especially students should not be wasted. If there is very little hope for replication or a particular project, then we have no business investigating the effectiveness of the project.

How about changes in system components not relevant to the project? Changes in components that are not relevant to the project will not affect the replicability, by definition. Nor will these changes affect the index, as the percent of variance accounted for is 0 and the contribution of that component would be 0. Unfortunately, in most educational systems, many components can influence the success of a project.

Implications

The implications of this article relate our position that statistical significance and effect size are important concepts, but they must be examined in light of replicability. Replicability is, in and of itself, not a one-dimensional concept but a multi-dimensional one. In this paper we identified two types of replication estimates. The first type is the SSER probability estimate, which is based on traditional statistical assumptions and probability concepts.

The second type is related to design and subjective probability issues. This approach provides a number of advantages. First, it can assist in the teaching of research design. That is, teaching this replication estimate emphasizes the need for researchers to attempt to identify the relevant variables in a study. Second, it can improve communication among researchers regarding relevant variables in a study in order to improve the design of such studies. Third, it encourages the use of meta-analysis to identify relevant variables. Fourth, it provides a method of stimulating the effects of the relevant variables on replicability of findings. One can stimulate small changes or large changes on relevant variables and the impact of these changes on replicability. As one can see, this second estimate is not a static approach but a dynamic one and may only be limited by the investigators' creativity and insight.

An emphasis on replications has implications for researchers regarding the research methodology they use. That is, researchers should consider conducting partial replications. Partial replication can be conducted by one or two approaches. First, half of the study could be an exact replication, and the other half could be an extension (into another grade level, using different in-service materials, or checking on efficacy in another bureaucratic situation). Second, the researcher could put a slight twist on the implementation, by reducing or eliminating a component, shortening the period, streamlining in-service, or monitoring more closely the actual implementation.

We believe that an emphasis on replicability estimates are as important to analyzing the data contained in a study as are the statistical test results and effect size estimates. The value of a study's results can be better assessed by researchers and practitioners when all three types of information (i.e., replicability estimates, statistical test results, and effect size estimates) are reported.

---

*Paper published in the *Mid-Western Educational Researcher,* (2004) Volume *17 (2)*, 36-40.

YOUR NOTES ON THE CHAPTERS

We have found that our readers like to write in notes, keep track of various ideas related to their learning. Use these pages to do that.

YOUR NOTES ON THE CHAPTERS

We have found that our readers like to write in notes, keep track of various ideas related to their learning. Use these pages to do that.

YOUR NOTES ON THE CHAPTERS

We have found that our readers like to write in notes, keep track of various ideas related to their learning. Use these pages to do that.

YOUR NOTES ON THE CHAPTERS

We have found that our readers like to write in notes, keep track of various ideas related to their learning. Use these pages to do that.

REFERENCES

Ahamann, J.S., & Glock, M.D. (1971). *Evaluating pupil growth: Principles of tests and measurements* (4th ed.). Boston: Allyn.

Asher, H.B. (1983). *Causal modeling* (2nd ed.). Sage University Paper Series on Quantitative Applications in the Social Science, 07-003. Beverly Hills, CA: Sage.

Benson, S.N.K., Aronson, E., Desmett, P., Shaheen, M., and Showalter, J. (2002). *Multiage education: A process and product evaluation.*_ Paper presented at the annual meeting of the Mid-western Educational Research Association, Columbus, OH

Bentler, P.M. (1989). *EQS structural equations program manual.* Los Angeles: BMDP Statistical Software.

Blalock, H.M. (ed.) (1970). *Causal models in the social sciences.* Chicago: Aldine Press.

Bray, J.H., & Maxwell, S.E. (1985). *Multivariate analysis of variance.* Sage University Paper Series on Quantitative Applications in the Social Science, 07-054. Beverly Hills, CA: Sage.

Campbell, D.T., and Stanley, J.C. (1963). *Experimental and quasi-experimental designs for research.* Chicago: Rand McNally College Publishing Co.

Cohen, J. (1968). *Statistical power analysis for the behavioral sciences.* New York: Academic Press.

Coleman, James S. (1966) *Equity of Educational opportunity.* ED012275

Cronbach, L.J. (1960). *Essentials of psychological testing* (2nd ed.). New York: Harper & Row Publishers.

(Deitchman, R., Newman, I., Burkholder, J., & Sanders, R. (1975). Type VI error: Inconsistency between the statistical procedure and the research question. Paper presented at the World Population Society Western Regional Meeting aboard Queen Mary, Long Beach, CA.

Downie, N.M., & Heath, R.W. (1959). *Basic statistical_methods.* New York: Harper & Brother Publishers.

Draper, N.R., & Smith, H. (1966). *Applied regression analysis.* New York: John Wiley and Sons

Dunteman, G.E. (1989). *Principal components analysis.* Sage University Paper Series on Quantitative Applications in the Social Science, 07-069. Beverly Hills, CA: Sage.

Ebel, R.L. (1965). *Measuring educational achievement.* New Jersey: Prentice Hall, Inc.

Educational Testing Service. (1965). *Short-cut statistics for teacher-made tests.* Princeton, NJ: Author.

Edwards, A.L. (1960). *Statistical methods of behavioral sciences.* New York: Rinehart & Co., Inc.

Edwards, A.L. (1972). *Experimental design in psychological research* (4th ed.). New York: Holt, Rinehart & Winston, Inc.

Ferguson, G.A. (1966). *Statistical analysis in psychology and education* (2nd ed.). New York: McGraw-Hill.

Flury, B. (1988). *Common principal components and related multivariate models.* New York: Wiley.

Fraas, J.W., and Newman, I. (2000). *Testing for statistical and practical significance: A suggested technique using a randomization test.* Paper presented at the annual meeting of the Mid-western Educational Research Association, Chicago, IL

Greenwalk, A.G., Gonzalez, R., Harris, R.J., and Guthrie, D. (1996). Effect sizes and p values: What should be reported and what should be replicated? Psychophysiology, 33, 175-183.

Harris, R.J. (1985). *A primer of multivariate statistics* (2nd ed.). Orlando, FL: Academic Press.

Joreskog, K.G., & Sorbom, D. (1988). *LISREL 7: A guide to the program and applications.* Chicago: SPSS.

Kaufman, L., & Rousseeuw, P.J. (1990). *Finding groups in data: An introduction to cluster analysis.* New York: Wiley.

Kelly, F., Newman, I., and McNeil, K. (1973). Suggested inferential statistical models for research and behavior modification. *Journal of Experimental Education, 4(4).*

Kerlinger, F.N. (1973). *Foundations of behavioral research* (2nd ed.). New York: Holt, Rinehart & Winston, Inc.

Klecka, W.R. (1980). *Discriminant analysis.* Sage University Paper Series on Quantitative Applications in the Social Science, 07-019. Beverly Hills, CA: Sage.

Klein and Newman (1974) Estimated parameters of three shrinkage estimated formuli. *Multiple Linear Regression Viewpoints, 4,* 6-11

Leuck and McGinn (1975). Regression analysis and educational production functions: Can they be trusted? *Harvard Review.*

Levin, J.R., and Robinson, D.H. (2000). Rejoinder: Statistical hypothesis testing, effect-size estimate, and the conclusion coherence of primary research studies. *Educational Researcher, 29(1),* 34-36.

Lomax, R.G. (1992). *Statistical concepts.* New York: Longman.

McNeil, K., Kelly, F.J. & McNeil, J.T. (1975). *Testing research hypotheses using multiple linear regression.* Carbondale, Southern Illinois University Press.

McNemar, Q. (1963). *Psychological statistics.* (3rd ed.). New York: John Wiley & Sons, Inc.

Mood, A.M. (1971). Partitioning variance in multiple regression analysis as a tool for developing learning models. American Educational Research Journal, 8(2), 191-202

Newman, I. (1973). Variations between shrinkage estimation formulas and the appropriateness of their interpretations. *Multiple Linear Regression Viewpoints.* Vo. 4, 45-48.

Newman, I. (1974). A demonstration of multiple regression models that will facilitate the investigation of trends and functional relationships. Paper presented at the Psychological Division of the Ohio Academy of Science, 83rd annual meeting, Wooster, OH.

Newman, I., Frye, B.J., & Newman, C. (1973). *An introduction to the basic concepts of measurement and evaluation.* Akron, OH: The University of Akron.

Nickerson, R.S. (2000). Null hypothesis significance testing: A review of an old and continuing controversy. *Psychological Methods, 5,* 241-301.

Popham, J.W. (1967). *Educational statistics: Use and interpretation.* New York: Harper & Row Publishers.

Posavac, E.J. (2002). Using *p* values to estimate the probability of a statistically significant Replications. *Understanding Statistics, 1(2),* 101-112.

Robinson, D.H., and Levin, J.R. (1997). Reflections on statistical and substantive significance, with a slice of replication. *Educational Researcher, 26(5),* 21-26.

Schlaffe, H. (1959). *The analysis of variance.* New York: Wiley

Swaminathan, H. (1989). Interpreting the results of multivariate analysis of variance. In B. Thompson (Ed.), *Advances in social science methodology* (Vol. 1, pp. 205-232). Greenwich, CT: JAI Press.

Tatsuoka, M.M. (1988). *Multivariate analysis* (2nd ed). New York: Macmillan.

Thompson, B. (1984). *Canonical correlation analysis: Uses and interpretation.* Sage University Paper Series on Quantitative Applications in the Social Science, 07-047. Beverly Hills, CA: Sage.

Thompson, B. (1996). AERA editorial policies regarding statistical significance testing: Three suggested reforms. *Educational Researcher, 25(2),* 26-30.

Thompson, B., (1997). Editorial policies regarding statistical significance tests: Further comments. *Educational Researcher, 26(5),* 29-32.

Thompson, B. (1999a). Statistical significance tests, effect size reporting and the vain pursuit of pseudo-objectivity. *Theory and Psychology, 9(2),* 191-196.

Thomson, B. (1999b). Journal editorial policies regarding statistical significance tests: Heat is to fire as *p* is to importance. *Educational Psychology Review, 11(2),* 157-169.

Williams, R.H. (1974). The effect of correlated errors of measurement on correlations among tests: A correlation for Spearman's correction for attenuation. *Journal of Experimental Education, 43(2),* 64-5.

Winer, B.J. (1971). *Statistical principles in experimental design.* New York: McGraw-Hill

APPENDIX A

A TABLE OF SUGGESTED TESTS OF SIGNIFICANCE FOR VARIOUS TYPES OF DATA AND SAMPLES

*Kind of Data	Single Group	Two Samples (Indepen-dent)	Two Samples Correlated (Depen-dent)	More Than Two Groups	
				Independent	Dependent
Nominal (Category or Frequency Data)	Chi Square (for a single sample) Fisher Exact Test	Chi Square (Multi-sample)	McNemar Tests	Chi Square (Multi-sample)	Cochran Q Test
Ordinal (Rank data)	Chi Square (for a single sample) Kilmogorov-Sminov	Chi Square (Multi-sample) Mann-Witney U Test	Matched pair signed ranks	Kruskal-Wallis Chi Square	Friedman Analysis of Variance
Interval or Ratio Data	Single Sample *t*-test	Indepen-dent Sample *t*-test	Dependent *t*-test (correlated *t*-test)	One-way F-test	Repeated Measures F-test

* The non-parametric tests are described in great detail in Siegel (1956). However, it is the authors' opinions (supported by much empirical data) that quite often the use of a non-parametric statistic is unnecessary and generally less powerful.

APPENDIX B

COMPUTATION OF A CORRELATION COEFFICIENT

AN EXAMPLE OF HOW TO COMPUTE A CORRELATION COEFFICIENT USING A CONCEPTUAL FORMULA

This appendix contains computational examples of:

A. How to calculate r using Z scores along with computations of the mean and standard deviations needed to compute the Z score.

B. The raw score formula for computing r.

Calculating r using Z scores

The formula for calculating r from Z scores is as follows:

$$r = \frac{\Sigma (Z_x)(Z_y)}{N-1}$$

where:

r = Pearson Correlation Coefficient

Σ = Summation

Z_x = Each score of x changed to a Z score

Z_y = Each score of y changed to a Z score

N = The Number of pair scores (x or y)

We will now use this formula to calculate the correlation for the following pairs of scores:

Subject	Score on X	Score on Y
1	10	8
2	6	5
3	4	5
4	8	9
5	7	11

The formula requires Z scores for both x and y; therefore, we will begin by calculating the Z scores for the data.
For x:

$$Z_x = \frac{x - \bar{x}}{S_x}$$

$$\bar{x} = \frac{\Sigma x}{N}$$

$$S_x = \sqrt{\frac{\Sigma (x - \bar{x})^2}{N-1}}$$

$$N = 5$$

$$\bar{x} = 35/5 = 7$$

Subject	Score on x	$(x - \bar{x})$	$(x - \bar{x})^2$	Z_x
1	10	+3	9	+3/2.24 = +1.34
2	6	-1	1	-1/2.24 = -.45
3	4	-3	9	-3/2.24 = -1.34
4	8	+1	1	+1/2.24 = + .45
5	7	0	0	0/2.24 = 0
Sum (Σ)	35	0	20	

$$S_x = \sqrt{\frac{\Sigma (x - \bar{x})^2}{N-1}} \quad \sqrt{\frac{20}{4}} = 2.24$$

For y:

$$Z_y = \frac{y - \bar{y}}{S_x}$$

$$\bar{y} = \frac{\Sigma \bar{y}}{N}$$

$$S_y = \sqrt{\frac{\Sigma (y - \bar{y})^2}{N-1}}$$

$$N = 5$$

$$\bar{y} = 385 = 7.6$$

Subject	Score on Y	$(y-\bar{y})$	$(y-\bar{y})^2$	Z_y
1	8	+.4	.16	.4/2.6 = .15
2	5	-2.6	6.76	-2.6/2.6 = -1
3	5	-2.6	6.76	-2.6/2.6 = -1
4	9	+1.4	1.96	1.4/2.6 = .54
5	11	+3.4	11.56	3.4/2.6 = 1.31
Sum (Σ)	38	0	20	

$$S = \sqrt{\frac{\Sigma (y - \bar{y})^2}{N-1}} = \sqrt{\frac{27.2}{4}} = 2.6$$

Now that we have calculated the Z scores, we can proceed to calculate the correlation between x and y.

$$r = \frac{\Sigma (Z_x)(Z_y)}{N-1}$$

Subject	Z_x	Z_y	$(Z_x)(Z_y)$
1	+1.34	.15	.20
2	-.45	-1	.45
3	-1.34	-1	1.34
4	.45	.54	.24
5	0	1.31	0
Sum (Σ)			2.23

$$r = \frac{2.23}{5-1} = .56$$

Note: If you already have the means and standard deviations and there are not too many scores, this procedure is relatively efficient.

Calculating r from raw scores

$$r = \frac{N \Sigma (x)(y) - (\Sigma x)(\Sigma y)}{\sqrt{[N \Sigma x^2 - (\Sigma x)^2][N \Sigma y^2 - (\Sigma y)^2]}}$$

Subjects	Score X	Score Y	(x)(Y)	X^2	Y^2
1	10	8	80	100	64
2	6	5	30	36	25
3	4	5	20	16	25
4	8	9	72	64	81
5	7	11	77	49	121
Sum (Σ)	35	38	279	265	316

$N = 5$

$\Sigma(x)(y) = 279$

$\Sigma(x)\ \Sigma(y) = (35)(38) = 1330$

$(\Sigma x)^2 = (35)^2 = 1225$

$(\Sigma y^2) = (38)^2 = 1444$

substituting into the raw score formula:

$$r = \frac{5(279) - (35)(38)}{\sqrt{[5(265) - (1225)][5(316) - 1444]}}$$

$$r = \frac{1395 - 1330}{\sqrt{[5(265) - (1225)][5(316) - 1444]}}$$

$$r = 65 / \sqrt{[(1325-1225)(1580-1444)]}$$

$$r = 65 / \sqrt{(100)(136)}$$

$$r = 65 / \sqrt{13600} = 65/116.62 = .56$$

APPENDIX C

COMPUTATION OF A CHI²

This appendix contains two computational examples of Chi²:

- A two class example with equal expectations per cell.
- A four group (class x gender) example with equal expectations per cell.

Example 1: Comparing Enrollment in Two Classes

$$\chi^2 = \frac{\Sigma (O-E)^2}{E}$$

Σ = Summation
O = Observed Value (Frequency)
E = Expected Value (Frequency)
df = (# of groups - 1)

Assume that we have 90 students with 70 enrolled in class 1 and 20 in class 2. We are interested in knowing if there is a significant difference in enrollment between the two classes (at $\alpha = .05$).

$O_1 = 70$ $E_1 = 45$ $O_2 = 20$ $E_2 = 45$

Class 1	Class 2
O_1 / E_1	O_2 / E_2

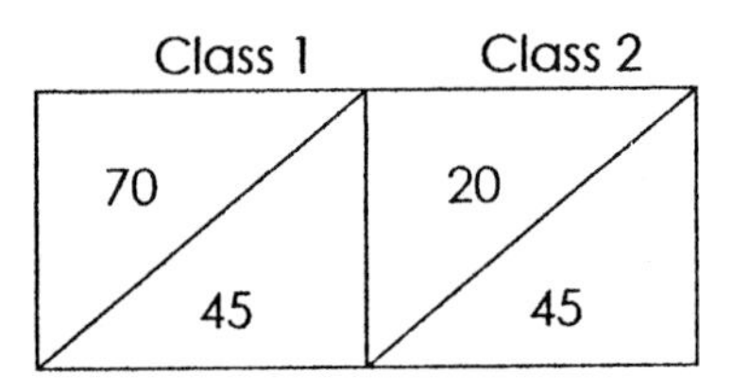

$$\chi^2 = \frac{\Sigma (O-E)^2}{E} = \frac{(70-45)^2 + (20-45)^2}{45} = \frac{625+625}{45}$$

$$\chi^2 = \frac{1250}{45} = 27.8 \quad df = 1 \quad \text{Significant}$$

There are significantly more students enrolled in class 1 than in class 2.

Example 2: Comparing 4 Groups (Class by Gender)

In this example, we have 2 classes (represented as columns) composed of both male and female students (represented as rows). We wish to know if there are significant differences in enrollment for males and females in these two classes.

$O_1 = 40$ $O_2 = 28$ $O_3 = 50$ $O_4 = 18$

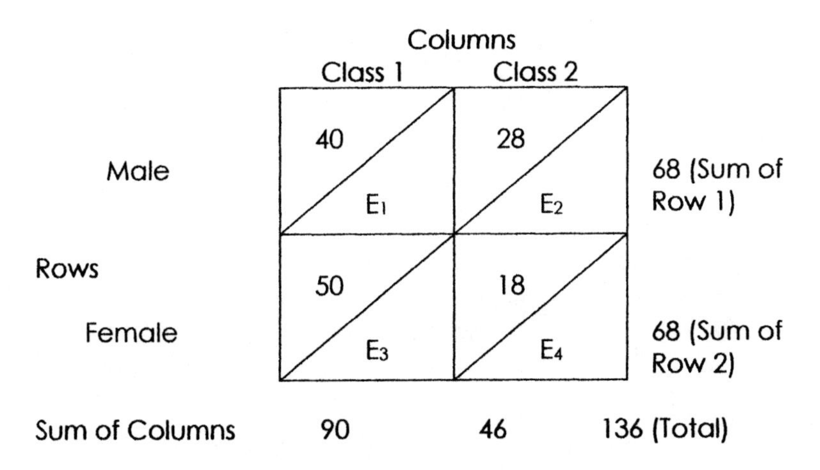

Columns
Class 1
Class 2
Male
40
E_1
28
E_2
68 (Sum of Row 1)
Rows
Female
50
E_3
18
E_4
68 (Sum of Row 2)
Sum of Columns
90
46
136 (Total)

To begin, we must first calculate the expected values for each cell. The expected values are calculated as follows:

$$E_1 = \frac{(\text{Row 1})*(\text{Column 1})}{\text{Total}} = (68*90)/136 = 45$$

$$E_2 = \frac{(\text{Row 1})*(\text{Column 2})}{\text{Total}} = (68*46)/136 = 23$$

$$E_3 = \frac{(\text{Row 2})*(\text{Column 1})}{\text{Total}} = (68*90)/136 = 45$$

$$E_4 = \frac{(\text{Row 2})*(\text{Column 2})}{\text{Total}} = (68*46)/136 = 23$$

Therefore, the completed table with expected values would be:

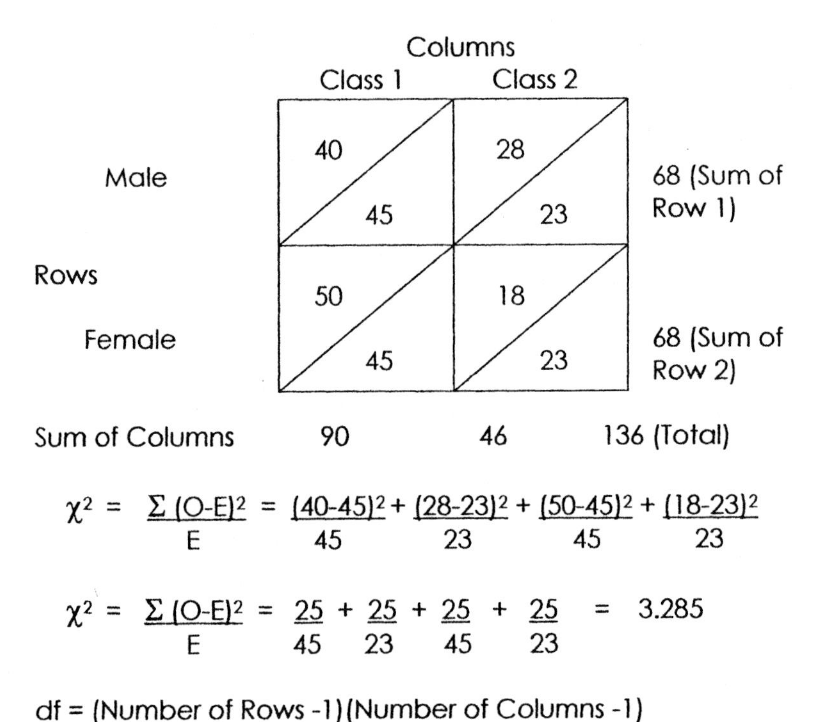

$$\chi^2 = \frac{\Sigma (O-E)^2}{E} = \frac{(40-45)^2}{45} + \frac{(28-23)^2}{23} + \frac{(50-45)^2}{45} + \frac{(18-23)^2}{23}$$

$$\chi^2 = \frac{\Sigma (O-E)^2}{E} = \frac{25}{45} + \frac{25}{23} + \frac{25}{45} + \frac{25}{23} = 3.285$$

df = (Number of Rows -1)(Number of Columns -1)
= (2-1)(2-1) = 1

$\chi^2_{critical}$ (α = .05 and 1 degree of freedom) = 3.841 which is greater than the obtained χ^2 value of 3.285; therefore, we conclude that the differences are not greater than would be expected by chance alone.

APPENDIX D

COMPUTATION OF AN INDEPENDENT *t*-TEST

AN EXAMPLE OF HOW TO COMPUTE A *t*-TEST

This appendix contains a computational example of a *t*-test for independent samples.

Given the following data:

Control Subjects	Group Scores
1	5
2	7
3	9
4	9
5	10
Sum	40

Experimental Subjects	Group Scores
6	0
7	2
8	3
9	4
10	6
Sum	15

$N_c = N_{(control)} = 5$
$\bar{X}_c = \bar{X}_{(control)} = 8$
$S_c = S_{(control)} = 2$

$N_e = N_{(experiment)} = 5$
$\bar{X}_e = \bar{X}_{(experiment)} = 3$
$S_c = S_{(experiment)} = 2.24$

$$t = \frac{\bar{X}_{(control)} - \bar{X}_{(experiment)}}{S_{D\bar{x}}}$$

where $S_{\bar{x}}$ = the estimated standard error of the difference between the means:

$$S_{D\bar{x}} = \sqrt{S_c^2 + S_e^2}$$

where $S_c^2 = \frac{S_{(control)}}{\sqrt{N_{(control)}}}$ and $S_e^2 = \frac{S_{(experiment)}}{\sqrt{N_{(experiment)}}}$

Therefore:

$$S_C = 2 / \sqrt{5} = 2/2.24 = .894$$

$$S_\varepsilon = 2.24 / \sqrt{5} = 2.24/2.24 = 1$$

$$S_C{}^2 = (.894)^2 = .80$$

$$S_\varepsilon{}^2 = (1)^2 = 1$$

$$S_{D\bar{X}} = \sqrt{.80 + 1} = \sqrt{1.8} = 1.341$$

$$t = \frac{\bar{X}_{(control)} - \bar{X}_{(experiment)}}{S_{D\bar{X}}} = \frac{8-3}{1.34} = 5/1.341 = 3.728$$

$t = 3.728$ df = (n-2) = 8
$t_{critical}$ ($\alpha = .05$, df = 8, two-tailed) = 2.306

The t value of 3.728 is greater than the critical value; therefore, we conclude that the differences between the scores of the experimental and control groups were greater than would be expected by chance alone. The experimental group scored lower than the control.

APPENDIX E

COMPUTATION OF A DEPENDENT *t*-TEST

AN EXAMPLE OF HOW TO COMPUTE A *t*-TEST FOR DEPENDENT (CORRELATIONAL) SAMPLES.

Given the following data:

Subjects	Test #1	Test #2	d	d^2
1	25	16	9	81
2	25	11	14	196
3	23	13	10	100
4	20	14	6	36
5	20	7	13	169
6	17	30	-13	169
7	16	8	8	64
8	14	15	-1	1
9	13	5	8	64
10	11	6	5	25
		Σ	59	905

N = 10 (the number of paired scores)
d = difference between scores
df = 9 (the number of pairs -1)

$$t = \sqrt{\frac{N-1}{[N\Sigma d^2/(\Sigma d)^2]-1}} = \sqrt{\frac{10-1}{[10(905)/(59)^2]-1}}$$

$$= \sqrt{\frac{9}{[9050/(3481)]-1}} = \sqrt{\frac{9}{[2.6]-1}} = \sqrt{\frac{9}{1.6}}$$

$t = 2.37$ df = $(n-1) = 9$ significant at $\alpha = .05$ ($t_{critical} = 2.262$)

APPENDIX F

COMPUTATION OF A ONE-WAY ANALYSIS OF VARIANCE

AN EXAMPLE OF HOW TO COMPUTE A ONE-WAY ANOVA

Treatment #1	Treatment #2	Treatment #3

Total SS = Between SS + Within SS

$$\text{Total SS} = \Sigma X^2_{total} - \frac{(\Sigma X_{total})^2}{N}$$

$$\text{Between SS} = \left(\frac{(\Sigma X_{treatment\ 1})^2}{n_1} + \frac{(\Sigma X_{treatment\ 2})^2}{n_2} + \frac{(\Sigma X_{treatment\ 3})^2}{n_3}\right) - \frac{(\Sigma X_{total})^2}{N}$$

$$\text{Within SS} = \Sigma X^2 - \left(\frac{(\Sigma X_{treatment\ 1})^2}{n_1} + \frac{(\Sigma X_{treatment\ 2})^2}{n_2} + \frac{(\Sigma X_{treatment\ 3})^2}{n_3}\right)$$

Treatment #1	
X_1	X_1^2
9	81
12	144
12	144
13	169
14	196
$\Sigma X_1 =$ 60	$\Sigma X_1^2 =$ 734

Treatment #2	
X_2	X_2^2
6	36
8	64
9	81
11	121
11	121
ΣX_2 = 45	ΣX_2^2 = 423

Treatment #3	
X_3	X_3^3
4	16
4	16
5	25
9	81
8	64
ΣX_3 = 30	$\Sigma X_3^2 =$ 202

$\Sigma X_{total} = \Sigma X_1 + \Sigma X_2 + \Sigma X_3 = 60 + 45 + 30 = 135$

$\Sigma X^2_{total} = \Sigma X_1^2 + \Sigma X_2^2 + \Sigma X_3^2 = 734 + 423 + 202 = 1359$

$$\text{Total SS} = \Sigma X^2_{total} - \frac{(\Sigma X_{total})^2}{N} = 1359 - \frac{(135)^2}{15}$$

$$= 1359 - \frac{18225}{15} = 1359 - 1215 = 144 = \text{Total SS}$$

$$\text{Between SS} = \left(\frac{(\Sigma X_{treatment\ 1})^2}{n_1} + \frac{(\Sigma X_{treatment\ 2})^2}{n_2} + \frac{(\Sigma X_{treatment\ 3})^2}{n_3}\right) - \frac{(\Sigma X_{total})^2}{N}$$

$$= \left(\frac{(60)^2}{5} + \frac{(45)^2}{5} + \frac{(30)^2}{5}\right) - \frac{(135)^2}{15}$$

$$= \left(\frac{3600}{5} + \frac{2025}{5} + \frac{900}{5}\right) - \frac{18225}{15}$$

$$= (720 + 405 + 180) - 1215 = 1305 - 1215 = 90$$

$$\text{Within SS} = \Sigma X^2 - \left(\frac{(\Sigma X_{treatment\ 1})^2}{n_1} + \frac{(\Sigma X_{treatment\ 2})^2}{n_2} + \frac{(\Sigma X_{treatment\ 3})^2}{n_3}\right)$$

$$= 1359 - 1305 = 54$$

This can also be found with the following method:
Within SS = Total SS – Between SS = 144 – 90 = 54

Source	SS	df	ms	F	Sign.
Between Treatments	90	k-1 = 3 - 1 = 2 k = # of groups	BSS/dfb= 90/2 = 45	ms_b/ms_w = 45/4.5 = 10	Yes p= .003
Within Treatments	54	N-K = 15 – 3 = 12 N= # of subjects	WSS/dfw = 54/12 = 4.5		
Total	144	N -1 = 14			

Note: $F_{critical}$ (α = .05, df numerator = 2, df denominator = 12) = 3.88

Please see Appendix R for a brief discussion and example of a correction for multiple comparisons.

APPENDIX G

COMPUTATION OF A TWO-WAY ANALYSIS OF VARIANCE

(3x3 FACTORIAL DESIGN)

AN EXAMPLE OF HOW TO COMPUTE A TWO-WAY ANOVA

In the following study, a researcher wishes to compare the effects of three teaching methods on achievement. Prior research has suggested that the methods may have a differential effect based on the IQ of the student; therefore, the researcher includes IQ as a second factor in the design. The factorial design is illustrated below:

		Methods (A)			
		Method 1	Method 2	Method 3	
	Low	Low IQ Method 1	Low IQ Method 2	Low IQ Method 3	ΣRow_1
IQ(B)	Med	Med IQ Method 1	Med IQ Method 2	Medium IQ Method3	ΣRow_2
	Hi	Hi IQ Method 1	Hi IQ Method 2	Hi IQ Method 3	ΣRow_3
		$\Sigma Column_1$	$\Sigma Column_2$	$\Sigma Column_3$	$\Sigma Total$

The researcher proceeded to collect data and enter it into the table as follows:

		Methods (A) Method 1	Method 2	Method 3	
IQ(B)	Low	9 12 12 13 14 $\Sigma Cell_1$ = 60	5 8 9 11 12 $\Sigma Cell_2$= 45	9 4 5 9 8 $\Sigma Cell_3$= 35	ΣRow_1 = 140
	Med	8 7 10 11 14 $\Sigma Cell_4$= 50	12 12 13 13 15 $\Sigma Cell_5$= 65	7 6 9 10 13 $\Sigma Cell_6$= 45	ΣRow_2 = 160
	Hi	5 5 6 10 9 $\Sigma Cell_7$= 35	6 8 9 10 12 $\Sigma Cell_8$= 45	9 11 11 12 12 $\Sigma Cell_9$= 55	ΣRow_3 = 135
		ΣCol_1 = 145	ΣCol_2 = 155	ΣCol_3 = 135	ΣTot = 435

$$\text{Total SS} = \Sigma X^2_{total} - \frac{(\Sigma X_{total})^2}{N_{total}}$$

$$SS_A \text{ (Column Effects)} = \left(\frac{(\Sigma Col_1)^2 + (\Sigma Col_2)^2 + (\Sigma Col_3)^2}{n_{per\ column}} \right) - \frac{(\Sigma X_{total})^2}{N_{total}}$$

$$SS_B \text{ (Row Effects)} = \left(\frac{(\Sigma Row_1)^2 + (\Sigma Row_2)^2 + (\Sigma Row_3)^2}{n_{per\ row}} \right) - \frac{(\Sigma X_{total})^2}{N_{total}}$$

$$SS_{AxB} \text{ (RowXColumn)} = \left(\frac{(\Sigma Cell_1)^2 + (\Sigma Cell)^2 + \ldots (\Sigma Cell_9)^2}{n_{per\ cell}} \right) - \frac{(\Sigma X_{total})^2}{N_{total}} - SS_A - SS_B$$

$$\text{Within SS} = \Sigma X^2_{total} - \left(\frac{(\Sigma Cell_1)^2 + (\Sigma Cell)^2 + \ldots (\Sigma Cell_9)^2}{n_{per\ cell}} \right)$$

						Sum of X	Sum of X^2
$Cell_1$	9	12	12	13	14	60	
$(C_1)^2$	81	144	144	169	196		734
$Cell_2$	5	8	9	11	12	45	
$(C_2)^2$	25	64	81	121	144		435
$Cell_3$	9	4	5	9	8	35	
$(C_3)^2$	81	16	25	81	64		267
$Cell_4$	8	7	10	11	14	50	
$(C_4)^2$	64	49	100	121	196		530
$Cell_5$	12	12	13	13	15	65	
$(C_5)^2$	144	144	169	169	22		851
$Cell_6$	7	6	9	10	13	45	
$(C_6)^2$	49	36	81	100	169		435
$Cell_7$	5	5	6	10	9	35	
$(C_7)^2$	25	25	36	100	81		267
$Cell_8$	6	8	9	10	12	45	
$(C_8)^2$	36	64	81	100	144		425
$Cell_9$	9	11	11	12	12	55	
$(C_9)^2$	81	121	121	144	144		611

ΣX_{total} = 435

ΣX^2_{total} = 4555

$$\text{Total SS} = \Sigma X^2_{total} - \frac{(\Sigma X_{total})^2}{N_{total}} = 4555 - \frac{(435)^2}{45}$$

$$= 4555 - \frac{(189225)}{45} = 4555 - 4205 = 350$$

$$SS_{within} = \Sigma X^2_{total} - \left(\frac{(\Sigma Cell_1)^2 + (\Sigma Cell)^2 + \ldots (\Sigma Cell_9)^2}{n_{per\ cell}} \right)$$

$$= 4555 - \left(\frac{(60)^2 + (45)^2 + (35)^2 + (50)^2 + (65)^2 + (45)^2 + (35)^2 + (45)^2 + (55)^2}{5} \right)$$

$$= 4555 - \frac{(3600 + 2025 + 1225 + 2500 + 4225 + 2025 + 1225 + 2025 + 3025)}{5}$$

$$= 4555 - \frac{21875}{5} = 4555 - 4375 = 180$$

$$SS_A \text{ (Column Effects)} = \left(\frac{(\Sigma Col_1)^2 + (\Sigma Col_2)^2 + (\Sigma Col_3)^2}{n_{per\ column}}\right) - \frac{\Sigma X_{total})^2}{N_{total}}$$

$$= \left(\frac{(145)^2 + (155)^2 + (135)^2}{15}\right) - \frac{(435)^2}{45}$$

$$= \left(\frac{(21025 + 24025 + 18225)}{15}\right) - \frac{189225}{45}$$

$$= 63275/15 - 4205 = 4218.333 - 4205 = 13.333$$

$$SS_B \text{ (Row Effects)} = \left(\frac{(\Sigma Row_1)^2 + (\Sigma Row_2)^2 + (\Sigma Row_3)^2}{n_{per\ row}}\right) - \frac{(\Sigma X_{total})^2}{N_{total}}$$

$$= \left(\frac{(140)^2 + (160)^2 + (135)^2}{15}\right) - 4205$$

$$= \left(\frac{(19600 + 25600 + 18225)}{15}\right) - 4205$$

$$= 63425/15 - 4205 = 4228.333 - 4205 = 23.333$$

$$SS_{AxB} \text{ (RowXColumn)} = \left(\frac{(\Sigma Cell_1)^2 + (\Sigma Cell)^2 + \ldots (\Sigma Cell_9)^2}{n_{per\ cell}}\right) - \frac{(\Sigma X_{total})^2}{N_{total}} - SS_A - SS_B$$

$$= 4375 - 4205 - 13.333 - 23.333 = 133.333$$

Source	SS	df	mS	F	Sign.
A	13.333	(A-1) = 2	SS_A/df_A = 6.666	mS_A/mS_W = 1.333	No
B	23.333	(B-1) = 2	SS_B/df_B = 11.666	mS_B/mS_W = 2.333	No
AxB	133.333	(A-1)(B-1) = 4	SS_{AB}/df_{AB} = 33.333	mS_{AB}/mS_W = 6.666	yes
Within	180	N-K = 45 – 9 = 36 N= # of subjects K= # of Cells	SS_W/df_W = 5		
Total	350	N -1 = 44			

df_A = (A-1) = (levels of main effect A) -1 = 3 - 1 = 2

df_B = (B-1) = (levels of main effect B) – 1 = 3 – 1 = 2

df_{AB} = (A-1)(B-1) = (A-1)(B-1) = (3-1)(3-1) = 4

df_W = (N-K) =(# of subjects) – (# of cells) = 45 – 9 = 36

$mS_A = SS_A/df_A$ = 13.333/2 = 6.666

$mS_B = SS_B/df_B$ = 23.333/2 = 11.666

$mS_{AB} = SS_{AB}/df_{AB}$ = 133.333/4 = 33.333

$mS_W = SS_W/df_W$ = 180/5 = 5

F for A main effect = mS_A/ mS_W = 6.666/5 = 1.333

F for B main effect = mS_B/ mS_W = 11.666/5 = 2.333

F for AB interaction = mS_{AB}/ mS_W = 33.333/5 = 6.666

APPENDIX H

AN EXAMPLE OF A SAS COMPUTER SETUP FOR THE CORRELATION PROBLEM IN APPENDIX B

This is an example of the SAS programming to run the correlation example from Appendix B. The program has a number of comment statements (statements placed in the tables) to explain the program steps. These should not be included when the program is entered.

data example1;
input id **1** scorex **3-4** scorey **6-7**;

The first line provides command (**data**) that names the data set. The second line (**input**) tells the program the names of the variables the column(s) in which each variable is located. Each command line is ended with a semicolon.

```
cards;
1 10 8
2 6 5
3 4 5
4 8 9
5 7 11
;
```

This is the data set. The data begins with a command (**cards**) which is followed by the data. Column number one contains the variable id, columns 3-4 contain the variable scorex, etc. The data concludes with a semicolon.

```
proc corr;
    title 'correlation between x and y';
    var scorex scorey;
run;
```

These are the procedure (**proc**). lines. The command (**corr**) runs a correlation between the variables specified in the **var** statement. All lines are concluded with a semicolon and program concludes with a **run;** statement.

Note: This is the output from the SAS program. This can be compared with the hand calculations from Appendix B.

The CORR Procedure
2 Variables: scorex scorey

Simple Statistics

Variable	N	Mean	Std Dev	Sum	Minimum	Maximum
scorex	5	7.00000	2.23607	35.00000	4.00000	10.00000
scorey	5	7.60000	2.60768	38.00000	5.00000	11.00000

These are the means and standard deviations for the variables x and y.

Pearson Correlation Coefficients, N = 5
Prob > |r| under H0: Rho=

	scorex	scorey
scorex	1.00000	0.55737 0.3290
scorey	0.55737 0.3290	1.00000

This is the correlation between x and y. The line below the correlation provides the probability value for the observed correlation.

APPENDIX I

AN EXAMPLE OF A SAS COMPUTER SETUP FOR THE CHI-SQUARE PROBLEM IN APPENDIX C

This is an example of the SAS programming to run the Chi-square example from Appendix C. The program has a number of comment statements (statements placed in the table) to explain the program steps. These should be deleted when the program is entered.

data example2;
input sex $ **1** class **3**;

The second line (**input**) tells the program the names of the variables. The $ following the variable sex indicates that this variable is coded alphanumerically (with characters rather than with a number).

cards;
m 1
m 1
m 1
m 1
m 1
m 1
m 1
m 1
m 1
m 1
m 1
m 1
m 1
m 1
m 1
m 1
m 1

m 1
m 1
m 1
m 1
m 1
m 1
m 1
m 1
m 1
m 1
m 1
m 1
m 1
m 1
m 1
m 1
m 1
m 1
m 1
m 1
m 1
m 1
m 1
m 1
m 2
m 2
m 2
m 2
m 2
m 2
m 2
m 2
m 2
m 2
m 2
m 2
m 2
m 2
m 2
m 2
m 2
m 2
m 2
m 2
m 2
m 2
m 2

m 2
m 2
m 2
m 2
m 2
f 1
f 1
f 1
f 1
f 1
f 1
f 1
f 1
f 1
f 1
f 1
f 1
f 1
f 1
f 1
f 1
f 1
f 1
f 1
f 1
f 1
f 1
f 1
f 1
f 1
f 1
f 1
f 1
f 1
f 1
f 1
f 1
f 1
f 1
f 1
f 1
f 1
f 1
f 1
f 1
f 1
f 1

```
f 1
f 1
f 1
f 1
f 1
f 1
f 1
f 1
f 2
f 2
f 2
f 2
f 2
f 2
f 2
f 2
f 2
f 2
f 2
f 2
f 2
f 2
f 2
f 2
f 2
f 2
;
proc freq;
        tables sex*class / chisq;
run;
```

These are the procedure (**proc**). lines. The command (**freq**) produces a frequency count for each of the variables. The command (**tables**) produces the frequency output in a table format. **Chisqr** performs a chi-square on the observed frequencies in the table.

Note: This is the output from the SAS program. This can be compared with the hand calculations provided in Appendix C.

The SAS System
The FREQ Procedure
Table of sex by class

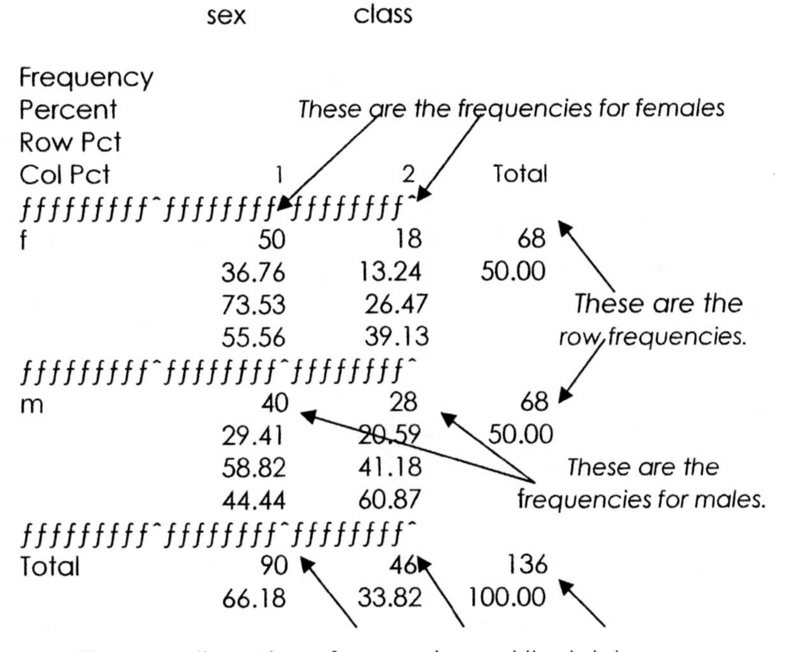

sex class

Frequency Percent Row Pct Col Pct	1	2	Total
f	50 36.76 73.53 55.56	18 13.24 26.47 39.13	68 50.00
m	40 29.41 58.82 44.44	28 20.59 41.18 60.87	68 50.00
Total	90 66.18	46 33.82	136 100.00

These are the frequencies for females
These are the row frequencies.
These are the frequencies for males.
These are the column frequencies and the total.

Statistics for Table of sex by class

Statistic	DF	Value	Prob
Chi-Square	1	3.2850	0.0699
Likelihood Ratio Chi-Square	1	3.3048	0.0691
Continuity Adj. Chi-Square	1	2.6609	0.1028
Mantel-Haenszel Chi-Square	1	3.2609	0.0710
Phi Coefficient		0.1554	
Contingency Coefficient		0.1536	
Cramer's V		0.1554	

This is the Chi-Square value and the probability of this observed value.

APPENDIX J

AN EXAMPLE OF A SAS COMPUTER SETUP FOR THE INDEPENDENT T-TEST PROBLEM IN APPENDIX D

This is an example of the SAS programming to run the independent t-test example from Appendix D. The program has a number of comment statements (statements placed in the table) to explain the program steps. These should not be included when the program is entered.

```
data example3;
input score 1-2 group 4;
cards;
05 0
07 0
09 0
10 0
00 1
02 1
03 1
04 1
06 1
;
```

This is the data set. The variable group is scored dichotomously (1 = experimental, 0 = control).

```
proc ttest;
    class group;
    var score;
run;
```

These are the procedure (**proc**). lines. The command (**ttest**) runs a t-test. The **Class** statement specifies the independent variable (group) with the two conditions to be compared. The **Var** statement specifies the dependent variable.

Note: This is the output of the SAS program. These results can be compared with the hand calculations in Appendix D.

The TTEST procedure

Statistics

This column contains the mean values for each group.

Experimental = 1, Control = 0

			Lower CL	Upper CL	Lower CL		Upper CL		
Variable	Class	N	Mean	Mean	Mean	Std Dev	Std Dev	Std Dev	Std Err
score	1	5	0.2236	3	5.7764	1.3397	2.2361	6.4255	1
score	0	5	5.5167	8	10.483	1.1983	2	5.7471	0.8944
score	Diff (1-2)		1.9062	5	8.0938	1.4329	2.1213	4.064	1.3416

These are the standard errors of the mean for each group.

T-Tests

Variable	Method	Variances	DF	t Value	Pr > |t|
score	Pooled	Equal	8	3.73	0.0058
score	Satterthwaite	Unequal	7.9	3.73	0.0059

This is the t value when one assumes equal variances in each group. The last column provides the p value this observed t.

Equality of Variances

Variable	Method	Num DF	Den DF	F Value	Pr > F
score	Folded F	4	4	1.25	0.8340

APPENDIX K

AN EXAMPLE OF A SAS COMPUTER SETUP FOR THE DEPENDENT T-TEST PROBLEM IN APPENDIX E

This is an example of the SAS programming to run the dependent t-test example from Appendix E. The program has a number of comment statements (statements placed in the table.) to explain the program steps. These should not be included when the program is entered.

```
data example4;
input test1 1-2 test2 4-5;
diff = test1 - test2;
```

This line creates a new variable (diff) to represent the difference between test1 and test2.

```
cards;
25 16
25 11
23 13
20 14
20 07
17 30
16 08
14 15
13 05
11 06
;
proc means
        N Mean stderr t prt;
        var diff;
run;
```

These are the procedure (**proc**). lines. The command (**means**) produces the mean for the variables specified in the **var** statement (diff). The commands following **Means (N, Mean, stderr, t, prt)** tell SAS to provide the sample size (N), mean, standard error, t value, and probability of the observed t.

Note: *This is the output from SAS. Compare this to the hand calculations in Appendix E.*

The SAS System
The MEANS Procedure
Analysis Variable : diff

N	Mean	Std Error	t Value	Pr > |t|
10	5.9000000	2.4875244	2.37	0.0418

APPENDIX L

AN EXAMPLE OF A SAS COMPUTER SETUP FOR THE ONE-WAY ANOVA PROBLEM IN APPENDIX F

This is an example of the SAS programming to run the one-way ANOVA example from Appendix F. The program has a number of comment statements (statements placed in the table) to explain the program steps. These should not be included when the program is entered.

```
data example5;
input group $1 score 3-4;
cards;
a 09
a 12
a 12
a 13
a 14
b 06
b 08
b 09
b 11
b 11
c 04
c 04
c 05
c 09
c 08
;
proc anova;
    class group;
    model score = group;
    means group;
run;
```

These are the procedure (**proc**). lines. The command (**anova**) runs an analysis of variance. The independent variable is specified in the **class** statement. In **proc anova**, the class variable must be categorical (a, b, c, or d) rather than continuous.
The **model** statement identifies the dependent variable (score) and specifies the independent variable(s) from which it is to be modeled. The final command (**means**) will produce the means for each of the classes (or categories of the independent variable).

Note: This is the first page of the SAS output. It simply provides information about the levels of the independent variable and the total sample size.

The ANOVA Procedure

Class Level Information

Class	Levels	Values
group	3	a b c

Number of observations 15

The ANOVA Procedure

Dependent Variable: score

These are the Between SS and the Mean Square Between.

Source	DF	Sum of Squares	Mean Square	F Value	Pr > F
Model	2	90.0000000	45.0000000	10.00	0.0028
Error	12	54.0000000	4.5000000		
Corrected Total	14	144.0000000			

This is the F ratio and the probability of the observed F.

This is the Total SS.

These are the Within SS and the Mean Square Within

R-Square	Coeff Var	Root MSE	score Mean
0.625000	23.57023	2.121320	9.000000

This is the proportion of variance in scores that can be accounted for by group membership (62 Y%).

These are the means and standard deviations of each of the subgroups.

The ANOVA Procedure

Level of group	N	------------score------------ Mean	Std Dev
a	5	12.0000000	1.87082869
b	5	9.0000000	2.12132034
c	5	6.0000000	2.34520788

APPENDIX M

AN EXAMPLE OF A SAS COMPUTER SETUP FOR THE TWO-WAY ANOVA PROBLEM IN APPENDIX G

This is an example of the SAS programming to run the two-way ANOVA example from Appendix G. The program has a number of comment statements (statements placed in the table) to explain the program steps. These should not be included when the program is entered.

```
proc format;
    value $iq   '1' = 'low' '2' = 'medium' '3' = 'high';
```

This is the first step in the process of placing value labels on the numeric values of a variable - this will tell SAS to label iq score of 1 as low, 2 as medium, and 3 as high.

```
run;


data example5;
input method 1 iq $ 3 score 5-6;
format iq $iq.;
```

This is the second step in the process of placing value labels on the numeric values of iq

```
cards;
1 1 09
1 1 12
1 1 12
1 1 13
1 1 14
```

2 1 05
2 1 08
2 1 09
2 1 11
2 1 12
3 1 09
3 1 04
3 1 05
3 1 09
3 1 08
1 2 08
1 2 07
1 2 10
1 2 11
1 2 14
2 2 12
2 2 12
2 2 13
2 2 13
2 2 15
3 2 07
3 2 06
3 2 09
3 2 10
3 2 13
1 3 05
1 3 05
1 3 06
1 3 10
1 3 09
2 3 06
2 3 08
2 3 09
2 3 10
2 3 12
3 3 09
3 3 11
3 3 11

```
3 3 12
3 3 12
;
proc anova;
    title '2 way ANOVA (3x3 design)';
    class method iq;
    model score = method | iq;
    means method | iq;
run;
```

These are the procedure (**proc**) lines. The independent variables are specified in the **class** statement. The | between method and iq tells SAS to create the interaction term for
the model (method*iq).

Note: This is the first page of the SAS output. It simply provides information about the levels of the independent variables and the total sample size.

The ANOVA Procedure

Class Level Information

Class	Levels	Values
method	3	1 2 3
iq	3	high low medium

Number of observations 45

2 way ANOVA (3x3 design)

The ANOVA Procedure

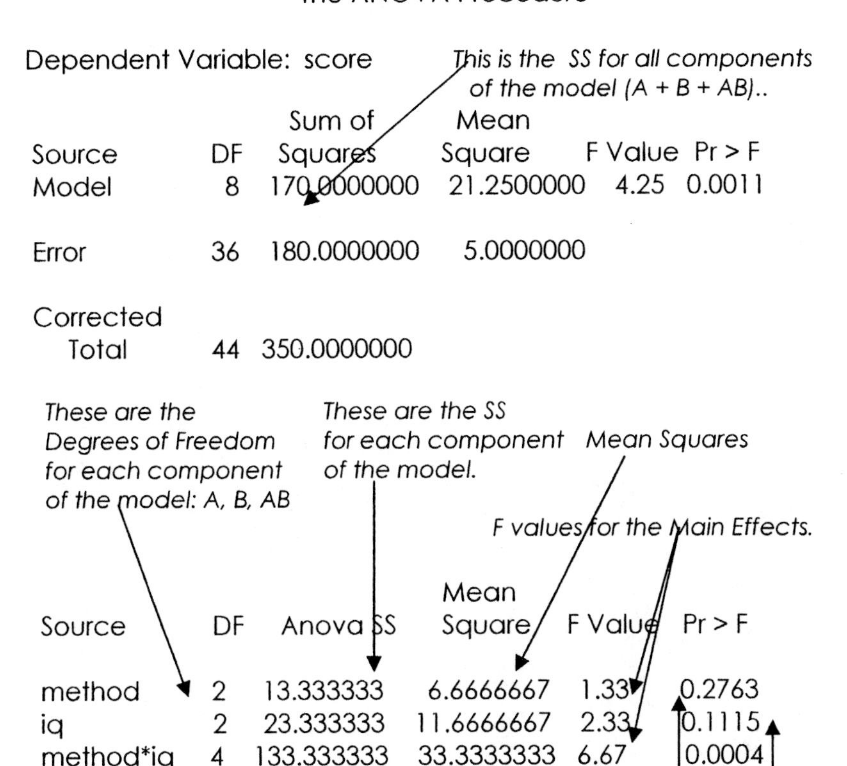

Dependent Variable: score

Source	DF	Sum of Squares	Mean Square	F Value	Pr > F
Model	8	170.0000000	21.2500000	4.25	0.0011
Error	36	180.0000000	5.0000000		
Corrected Total	44	350.0000000			

Source	DF	Anova SS	Mean Square	F Value	Pr > F
method	2	13.333333	6.6666667	1.33	0.2763
iq	2	23.333333	11.6666667	2.33	0.1115
method*iq	4	133.333333	33.3333333	6.67	0.0004

This is the SS for all components of the model (A + B + AB)..
These are the Degrees of Freedom for each component of the model: A, B, AB
These are the SS for each component of the model.
Mean Squares
F values for the Main Effects.
F value for the interaction term
Probabilities

These are the means and standard deviations of each of the subgroups.

2 way ANOVA (3x3 design)

Level of method	N	------------score------------ Mean	Std Dev
1	15	9.6666667	3.03942350
2	15	10.3333333	2.79455252
3	15	9.0000000	2.64575131

Level of iq	N	------------score------------ Mean	Std Dev
high	15	9.0000000	2.50713268
low	15	9.3333333	3.03942350
medium	15	10.6666667	2.79455252

Level of method	Level of iq	N	------------score------------ Mean	Std Dev
1	high	5	7.0000000	2.34520788
1	low	5	12.0000000	1.87082869
1	med	5	10.0000000	2.73861279
2	high	5	9.0000000	2.23606798
2	low	5	9.0000000	2.73861279
2	med	5	13.0000000	1.22474487
3	high	5	11.0000000	1.22474487
3	low	5	7.0000000	2.34520788
3	med	5	9.0000000	2.73861279

APPENDIX N

GENERAL COMMENTS IN REGARD TO RUNNING THE EXAMPLE PROBLEMS USING SPSSS

This appendix contains general information which is necessary to be better understand the SPSS examples that follow in Appendices O- .

I. SPSS, like SAS, operates with three windows: (1) the data window, (2) the output window, and (3) the syntax window. Each of these windows can be opened through the **FILE** menu.

II. The data window in SPSS is in spreadsheet format. The top of each column (in the data view) provides the name of the variable in that column. The left-most column provides the case (subject) numbers, and each row represents a subject (case). Data sets for each example will be given in a table format which can be copied directly into the spreadsheet format of SPSS.

III. All the following examples can be run from the data window through the use of pull-down menus. The instructions for using the pull-down menus will identify the menu selections sequentially. This will follow this general format:

Example for running descriptive statistics - –

select -> Analyze (*first menu*)
  select -> Descriptive Statistics (*second menu*)
    select -> Descriptives (*third menu*)

In the above example, one would first click on Analyze. This would bring up a menu with the option for Descriptive Statistics. One would then click on Descriptive Statistics bringing up yet another menu with the option for Descriptives. Clicking on Descriptives brings up a screen used to select the variables for the analysis.

IV. Analyses can also be run directly from the syntax window. The syntax necessary for running each example will also be provided. If you are attempting to run the examples in SPSS using this syntax, it will be necessary to ensure that the variable names are identical to the names used in the syntax examples provided. If the syntax names do not match variable names in your data set, SPSS will not recognize the variable name in the syntax; and, therefore, will not perform the analysis.

APPENDIX O

SPSS ANALYSIS OF THE CORRELATION EXAMPLE IN APPENDIX B

I. This is the data used in the example problem from Appendix B:

scorex	Scorey
10.00	8.00
6.00	5.00
4.00	5.00
8.00	9.00
7.00	11.00

II. These are the steps necessary to run the correlation from the data window:

select -> Analyze (*first menu*)
select -> Correlate (*second menu*)
select -> Bivariate (*third menu*)

This will bring up a window with options.

Highlight scorex and scorey and click the arrow which will place scorex and scorey in the "**Variables**" box.

Click "**OK**"

III. This is the syntax to do the above analysis:

```
CORRELATIONS
 /VARIABLES=scorex scorey
 /PRINT=TWOTAIL NOSIG
 /MISSING=PAIRWISE .
```

IV. This is the SPSS output:

Correlations

		SCOREX	SCOREY
SCOREX	Pearson Correlation	1	.557
	Sig. (2-tailed)	.	.329
	N	5	5
SCOREY	Pearson Correlation	.557	1
	Sig. (2-tailed)	.329	.
	N	5	5

APPENDIX P

SPSS ANALYSIS OF THE CHI-SQUARE EXAMPLE IN APPENDIX C

I. This is the data (partial – like cases have been omitted for the sake of space) used in the example problem from Appendix C. Where Gender =1 is male and Gender = 0 is female and class =1 is class #1 and class = 0 is class #2:

Subject	Gender	Class
1	1	1
2	1	1
..	..	..
40	1	1
41	1	0
42	1	0
..	..	..
68	1	0
69	0	1
70	0	1
..	..	..
118	0	1
119	0	0
120	0	0
..	..	..
136	0	0

Subjects 1-40 were males (1) and were in class #1 (1)
Subjects 41-68 were males (1) and were in class #2 (0)
Subjects 69-118 were females (0) and were in class #1 (1)
Subjects 119-136 were females (0) and were in class #2 (0)

Note: It is necessary to enter all of the cases into SPSS in order to perform the analyses!

II. These are the steps necessary to run the correlation from the data window:

select -> Analyze (*first menu*)
select -> Correlate (*second menu*)
select -> Crosstabs (*third menu*)

This will bring up a window with options.

Place gender in the box called "Rows" and place class in the box named "Columns."

Click on the button called "Statistics." This will open another window with options. Check Chi-Square.

Click "Continue" - This will close the options window.

Click "OK"

III. This is the syntax to do the above analysis:

```
CROSSTABS
 /TABLES=gender  BY class
 /FORMAT= AVALUE TABLES
 /STATISTIC=CHISQ
 /CELLS= COUNT .
```

IV. This is the SPSS output:

Crosstabs

Case Processing Summary

	Cases					
	Valid		Missing		Total	
	N	Percent	N	Percent	N	Percent
GENDER * CLASS	136	100.0%	0	.0%	136	100.0%

GENDER * CLASS Crosstabulation

Count

		CLASS		Total
		class 2	class 1	
GENDER	female	18	50	68
	male	28	40	68
Total		46	90	136

Chi-Square Tests

	Value	df	Asymp. Sig. (2-sided)	Exact Sig. (2-sided)	Exact Sig. (1-sided)
Pearson Chi-Square	3.285[b]	1	.070		
Continuity Correction[a]	2.661	1	.103		
Likelihood Ratio	3.305	1	.069		
Fisher's Exact Test				.102	.051
Linear-by-Linear Association	3.261	1	.071		
N of Valid Cases	136				

a. Computed only for a 2x2 table

b. 0 cells (.0%) have expected count less than 5. The minimum expected count is 23.00.

APPENDIX Q

SPSS ANALYSIS OF THE INDEPENDENT T-TEST EXAMPLE IN APPENDIX D

I. This is the data used in the example problem from Appendix B:

Group	Scores
1	5
1	7
1	9
1	9
1	10
0	0
0	2
0	3
0	4
0	6

Where: Group = 1 is control and Group = 0 is experimental

II. These are the steps necessary to run the correlation from the data window:

select -> Analyze (*first menu*)

select -> Compare Means (second *menu*)

select -> Independent Samples TTEST (*third menu*)

This will bring up a window with options.

Place score in the in the box called "Test Variable" and place group in the box named "Grouping Variable."

Click on the button called "Define Groups." Place a "1" in the box called "Group 1" and a "0" in the box called "Group 2."

Click "Continue"

Click "OK"

III. This is the syntax to do the above analysis:

```
T-TEST
  GROUPS=group(1 0)
  /MISSING=ANALYSIS
  /VARIABLES=score
  /CRITERIA=CIN(.95) .
```

IV. This is the SPSS output:

T-Test

Group Statistics

	GROUP	N	Mean	Std. Deviation	Std. Error Mean
SCORE	control	5	8.0000	2.00000	.89443
	experimental	5	3.0000	2.23607	1.00000

Independent Samples Test

		Levene's Test for Equality of Variances		t-test for Equality of Means						
									95% Confidence Interval of the Difference	
		F	Sig	t	df	Sig (2-tailed)	Mean Difference	Std. Error Difference	Lower	Upper
SCORE	Equal variances assumed	.000	1.000	3.727	8	.006	5.0000	1.34164	1.90617	8.09383
	Equal variances not assumed			3.727	7.902	.006	5.0000	1.34164	1.89951	8.10049

This is the t value when one assumes equal variances in each group.

This is the number of degrees of freedom for the t-test
This is the probability of the observed t value

APPENDIX R

SPSS ANALYSIS OF THE DEPENDENT T-TEST EXAMPLE IN APPENDIX E

I. This is the data used in the example problem from Appendix E:

Test1	Test2
25	16
25	11
23	13
20	14
20	07
17	30
16	08
14	15
13	05
11	06

II. These are the steps necessary to run the correlation from the data window:

select -> Analyze (*first menu*)
select -> Compare Means (*second menu*)
select -> Paired-Samples TTEST (*third menu*)

This will bring up a window with options.

Place test1 and test2 as a pair into the box labeled "Paired Variables."

Click "OK"

III. This is the syntax to do the above analysis:

```
T-TEST
  PAIRS= test1  WITH test2 (PAIRED)
  /CRITERIA=CIN(.95)
  /MISSING=ANALYSIS.
```

IV. This is the SPSS output:

T-Test

Paired Samples Statistics

		Mean	N	Std. Deviation	Std. Error Mean
Pair 1	TEST1	18.4000	10	4.99333	1.57903
	TEST2	12.5000	10	7.29155	2.30579

Paired Samples Correlations

		N	Correlation	Sig.
Pair 1	TEST1 & TEST2	10	.223	.536

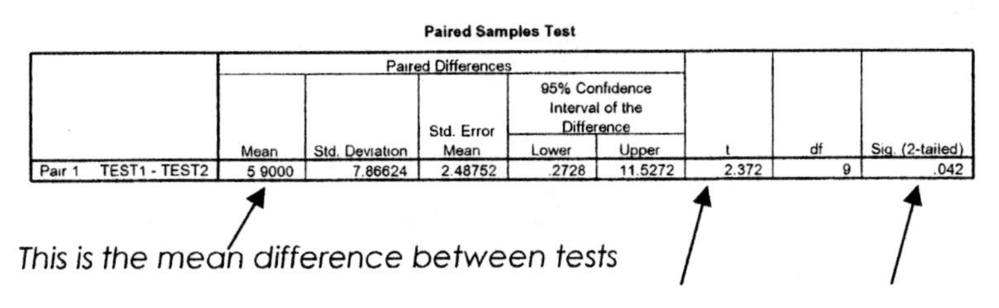

Paired Samples Test

		Paired Differences					t	df	Sig. (2-tailed)
					95% Confidence Interval of the Difference				
		Mean	Std. Deviation	Std. Error Mean	Lower	Upper			
Pair 1	TEST1 - TEST2	5.9000	7.86624	2.48752	.2728	11.5272	2.372	9	.042

This is the mean difference between tests

This is the t value and the probability of this observed t value

APPENDIX S

SPSS ANALYSIS OF THE ONE-WAY ANOVA EXAMPLE IN APPENDIX F

I. This is the data used in the example problem from Appendix F

Group	Score
1	09
1	12
1	12
1	13
1	14
2	06
2	08
2	09
2	11
2	11
3	04
3	04
3	05
3	09
3	08

II. These are the steps necessary to run the correlation from the data window:

select -> Analyze (*first menu*)
select -> Compare Means (second *menu*)
select -> One-Way ANOVA (*third menu*)

This will bring up a window with options.

Place Score into the box labeled "Dependent Variable."
Place Group into the box labeled "Factor."

Click "OK"

III. This is the syntax to do the above analysis:

```
ONEWAY
 score BY group
 /MISSING ANALYSIS .
```

IV. This is the SPSS output:

ANOVA

SCORE

	Sum of Squares	df	Mean Square	F	Sig.
Between Groups	90.000	2	45.000	10.000	.003
Within Groups	54.000	12	4.500		
Total	144.000	14			

APPENDIX T

SPSS ANALYSIS OF THE TWO-WAY ANOVA EXAMPLE IN APPENDIX G

I. This is the data used in the example problem from Appendix F

Method	IQ	Score		Method	IQ	Score
1	1	09		2	2	13
1	1	12		2	2	15
1	1	12		3	2	07
1	1	13		3	2	06
1	1	14		3	2	09
2	1	05		3	2	10
2	1	08		3	2	13
2	1	09		1	3	05
2	1	11		1	3	05
2	1	12		1	3	06
3	1	09		1	3	10
3	1	04		1	3	09
3	1	05		2	3	06
3	1	09		2	3	08
3	1	08		2	3	09
1	2	08		2	3	10
1	2	07		2	3	12
1	2	10		3	3	09
1	2	11		3	3	11
1	2	14		3	3	11
2	2	12		3	3	12
2	2	12		3	3	12
2	2	13				

Where IQ = 1 is low; IQ = 2 is medium; IQ = 3 is high

II. These are the steps to run the correlation:

select -> Analyze (*first menu*)
  select -> General Linear Model (second)
    select -> Univariate (*third menu*)

This will bring up a window with options. Place Score into the box labeled "Dependent Variable."
Place Method and IQ into the box labeled "Fixed Factor."
Click "OK"

III. This is the syntax to do the above analysis:

```
UNIANOVA
  score BY method iq
  /METHOD = SSTYPE(3)
  /INTERCEPT = INCLUDE
  /CRITERIA = ALPHA(.05)
  /DESIGN = method iq method*iq .
```

IV. This is the SPSS output:

Univariate Analysis of Variance

Between-Subjects Factors

		Value Label	N
METHOD	1.00		15
	2.00		15
	3.00		15
IQ	1.00	low	15
	2.00	medium	15
	3.00	high	15

Tests of Between-Subjects Effects

Dependent Variable: SCORE

Source	Type III Sum of Squares	df	Mean Square	F	Sig.
Corrected Model	170.000[a]	8	21.250	4.250	.001
Intercept	4205.000	1	4205.000	841.000	.000
METHOD	13.333	2	6.667	1.333	.276
IQ	23.333	2	11.667	2.333	.111
METHOD * IQ	133.333	4	33.333	6.667	.000
Error	180.000	36	5.000		
Total	4555.000	45			
Corrected Total	350.000	44			

a. R Squared = .486 (Adjusted R Squared = .371)

GLOSSARY

active variable - an independent variable that is under the control of and can be manipulated by the experimenter.

age norm - represents the average score received in a given test by a specific age group; typically reported on a table which depicts the average score on a given test by various age groups.

assigned variable - sometimes referred to as an attribute, it is an independent variable not under the control of the experimenter. Generally, sex, race, and intelligence are assigned variables.

average - typically used to refer to a measure of central tendency of a group of scores.

biased sampling - each person in the population did <u>not</u> have an equal chance of being selected for the sample.

canonical correlation - an extension of multiple regression analysis which predicts a set (2 or more) of dependent variables from a set (2 or more) of independent variables. This technique estimates the relationship between two sets of continuous variables--a predictor set and a criterion set. One can think of these two sets as principal components. Each set of principal components is weighted to maximize the prediction of the other set.

cluster analysis - a data reduction technique like principal components. The purpose is to group persons or objects into a smaller number of groups based upon their similarities, a classification procedure. Objects or persons who are in a cluster are more similar to each other than those from other clusters.

common factor variance - variability shared by two or more variables. It is related to the concept of correlation and covariance (see r^2).

concept - an abstraction, it is a generalization of specific qualities; for example, weight, height age.

confounding - the mixing of the variance of one or more independent variables with one or more extraneous variables.

construct - a concept with added meaning, such as I.Q.

content validity - a test which displays this characteristic would be typified by questions which are drawn from subject matter which is of concern and interest and has data to support that the items are representative of the subject matter.

continuous variables - a variable with many different values.

contrast (pair) comparisons - probability of making a Type I error for any specific comparison.

counterbalanced design - a design in which experimental precision is enhanced by having all treatments responded to by all subjects, but in a different order; for example, crossover design and Latin Square design.

covariance - the extent that two or more variables are correlated with each other and how much variability in the measurements can be attributed to the commonality between the variables.

covary - two or more variables are correlated and they vary systematically with other, either positively or negatively.

crosstabs - a procedure in which data can be tabulated, usually frequencies (nominal data) or percentages; it is used when analyzing the relationship between categorical or nominal data.

crossed-design - is a factorial design in which all levels of one factor can be found in all levels of the other factor or factors.

degrees of freedom - related to the total number of independent replications (n) which can be the subject, items, groups, etc.

dependent variable - is referred to as the criterion, the "then" part of the 'if-then' hypothesis statement; it's the inferred effect that is being measured due to the independent variable; it is what is always statistically analyzed.

dichotomous variable - a variable that has only two values assigned to it.

discriminant analysis - purpose is to predict group membership based on some weighted linear set of predictor variables.

disordinal interaction - interaction in which when the lines are plotted, they cross in the area of interest.

empirical - objective observation (or tests).

error of measurement - variance due to lack of reliability in the measurement instrument.

error variance - uncontrolled variance, considered to be a function of random variation in measures due to chance

eta - symbolized by η^2. Eta is only interpreted in terms of Eta^2. It is also called the correlation ratio. It ranges from a score of 0 (no relation) to a score of 1 (perfect relation). Eta measures the total degree of relationship both linear and non linear. That is, it measures straight line and curved line relationships ($\eta^2 yx = 1 - (SSwithin/SStotal)$)

evaluation – a process that represents the sum total of the measures of something within a given class.

experimental operational definition - defines a variable in terms of its procedures or operations, how the variable was manipulated.

experimental research - there must be at least one active variable under the experimenter's control.

experimental variance (between group) - variation in measures on the dependent variable attributed to systematic differences between our independent variables.

ex post facto research - the independent variable is not under the control of the researcher and/or has already occurred.

external validity - the extent that a study is generalizable to other people, groups, investigations, etc.

extraneous variables - variables that one does not want to include, but does want to control for possible contamination of the study.

face validity - a measurement instrument characteristic which indicates that the questions being asked either do or do not appear to represent the content of the instruction: this represents a visual examination of the items and instrument.

factor analysis - a multivariate method which helps the researcher to determine and interpret how many dimensions (factors) exist for a particular set of measures; this is done through a mathematical procedure of analyzing sets of correlations. A factor is a hypothetical construct that underlies a set of variables (items). A factor is made up of items that load on the factor (correlated with the factor). An item is considered to load on a factor if it is highly or relatively highly correlated with that factor. If the item is only highly correlated with one factor, it is called factor pure. A factor is interpreted based on the item that loads on the factor.

factorial design - a design made up of at least two factors (main effects), the purpose of which is to control variance while testing the effects of the independent variables in terms of different dependent variable scores.

family-wise comparisons - the probability of making a Type I error in a set of contrasts (family).

fixed design - all levels of a factor in a factorial design are fixed. That is, one is not allowed to generalize to any level other than the one that has been tested.

heterogeneity - this term refers to the degree of difference or dissimilarity of subjects.

heuristic - the research's potential value for further investigation.

homogeneity - this term refers to the degree of similarity or commonality among subjects.

hypothesis - a statement of the relationship between two or more variables in an if-then form.

independent variable - there are two types - see active and assigned variable; the presumed cause of the dependent variable; the 'if' part of the 'if-then' hypothesis statement.

interaction - the differential effect across the area of interest (it is the simple effects that are plotted; the lines will be nonparallel).

internal validity - a study has internal validity to the extent that one can say the independent variable causes the effects of the dependent variable; in other words, a study has this to the extent it can assume causation.

interquartile range - is an estimate of how much the scores in the distribution deviate from the measure of central tendency as measured by the median. It is the [75% - 25% divided by 2] sometimes also called quartile deviation.

interval - numbers which represent equal units of measurement; for example, distance between 2 and 4 is equal to the distance between 10 and 12; such scales can be added, subtracted, multiplied and divided; these scales do not have an absolute zero; for example, a thermometer.

intervening variable - sometimes called constructs, a term invented, that is assumed to account for unobservable internal psychological processes; it is generally inferred from behavior.

law of large numbers - as you increase the size of the sample, the more accurately your sample value is representative of the population value.

least square solution - is the general statistical computational procedure on which tests of significance are calculated. The least

square solution is a procedure in which the sum of all the deviation scores squared are as small as possible.

main effects - the factors that make up a factorial design, they must have at least two levels.

maxicon principle - maximize experimental variance while minimizing error variance.

mean - a measure of central tendency that is computed in the same manner as the arithmetic average; sum of all the scores divided by the number of scores.

measurement - this term refers to the assessment of student behavior in quantitative terms; an evaluation plan will normally encompass several measures of student behavior.

measured operational definition - defines a variable or construct in terms of how it is being measured.

median - this measure of central tendency indicates the middle score of a group of scores; it is a point in a distribution in which half of the scores fall below and the other half fall above that point.

metaphysical explanation - a relationship or proposition that cannot be empirically tested.

mixed design - a factorial analysis of variance design in which at least one factor is random and at least one factor is fixed.

mode - a measure of central tendency, the score which occurs most frequently.

multiple regression - a method for predicting a dependent variable by two or more independent variables that may be continuous, dichotomous or both.

multivariable - when there is one dependent variable being predicted by two or more independent variables.

multivariate - when there are two or more dependent variables being predicted by two or more independent variables.

nested design - is when all levels of one factor are <u>not</u> found in all levels of another. The levels of one factor are confounded in levels of another factor. Sometimes called split-plot design.

nominal - a measurement scale made up of numbers used to designate a class or category; they cannot be added, subtracted, multiplied, or divided; for example, use numbers to classify people by sex, religion, race, etc.

nonparametric statistics - statistics that don't make the same stringent assumptions as do parametric statistics; are also called distribution free tests (i.e., they are free of the assumptions about the distribution of the population and sample [normal, skewed, etc.]); do not make the assumption that requires the measurements to be at least interval in nature and also do not require homogeneity of variance. However, they tend to be less powerful than parametric statistics.

non-probability sampling - sampling that does not use random sampling procedures.

norm - this term is used to refer to typical scores obtained by relatively large groups of people with similar characteristics; a test may be "normed" by compiling typical scores for a certain grade level or age level or other common base.

normal distribution - this is a theoretical distribution of scores in which a plotted frequency distribution will result in a smooth bell-shaped curve. Such a distribution of scores will result in approximately 68% of the scores between ± one standard deviation; 95% of the scores between ± two standard deviations; and 99.7% of the scores between ± three standard deviations.

null hypothesis - states that there is no relationship between two or more variables.

objectivity - a method of measurement that is not affected by the researcher's biases.

operational definition - defines a variable or construct in terms of how it is being measured or its activities (procedures).

ordinal - scales that are rank ordered; numbers are assigned to represent rank ordered positions; they can not be added or subtracted, multiplied or divided; for example, size, places, IQ.

ordinal interaction - interaction when the lines are plotted, they do not cross in the area of interest, but the lines are not parallel.

orthogonal - variables are at right angles to each other, zero correlated with each other and independent of each other.

orthogonal comparisons - a set of contrasts that are independent (nonredundant) for K number of groups. Only K-1 orthogonal contrasts can be moded.

parameter - the actual population value.

parametric statistics - statistics that make certain assumptions about the population from which the sample comes. Assumes: normal distribution of the population and sample, homogeneity of variance, and equal intervals of its measures.

path analysis - method for studying the relationship -between direct and indirect effects assumed (causes) of the dependent variables (endogenous). It is a procedure used to test the causal relationship formulated by a theoretical model. (Theory is needed before one can effectively use path analysis.) Path analysis generally uses path analytic diagrams to display the pattern of the causal relationships among a set of predictor variable (exogenous) and dependent variable (endogenous). Path analysis is used to possibly better understand "causal relationships" when non-experimental conditions exist.

percentile rank - this is a method of expressing individual scores in terms of relative rank among the total group of scores; a particular

numerical percentile rank indicates the percentage of scores which fall below the given rank.

power - the power of a test is the ability of a test to detect a difference when a difference exists. Mathematically stated:

power = 1 - (the probability of making a Type II error)
(Type II errors are also called beta errors.)

practical significance - differs from statistical significance, deals with whether the difference is large enough to be useful, and sometimes is measured in terms of the proportion of variance accounted for (r^2).

principal components analysis (PC) - a special case of factor analysis. It is a data reduction procedure where variables (items, etc.) are transformed into a smaller set of linear combinations of these variables (items). These linear combinations are called principal components (factors) and these principal components are uncorrelated with one another. The first step in doing principal component analysis is to start with a correlation matrix. Conceptually, it creates PC by putting together the most highly correlated set of items, and each set of highly correlated items is calculated in such a way to create a zero correlation with every other set.

probability sampling - uses random sampling procedures.

problem - a question statement asking what relationship exists between two or more variables; many hypotheses can be derived from a problem statement.

quasi-experimental research - there is an active variable but the investigator does not have total control over scheduling or can't randomly assign.

qualitative measures - assessments of this type are more subjective in that one is attempting to assess levels of quality instead of quantity rank.

quantitative measures - measures of this type are more objective in nature in that weights or values are assigned for specific levels of a characteristic or performance; such weights or values are typically expressed in a numerical form; measurement is virtually impossible to perform without quantitative values identified for a given characteristic or performance.

quota samples - a form of non-probabilistic sampling in which quotas or proportions of particular objects or groups are used for representatives; it appears similar to the probabilistic technique of stratified random sampling but it is generally not as accurate.

r^2 - also referred to as the coefficient of determination; the proportion of variance that one variable can account for in another variable (common factor variance).

random assignment - a procedure in which subjects or objects are assigned to different groups using random procedures.

randomization - assignment of subjects, objects, treatments, etc., in such a way that each of these has an equal chance of being assigned by using random procedure.

randomized design - the levels of a factor are randomly picked from a population for the purpose of generalizing from these levels to the "infinitely large" population from which they were picked.

random sample - method of sampling so that every person in the population being sampled has an equal chance of being drawn.

range - this term reflects a measure of variability of a group of scores in that it expresses the difference between the highest and lowest scores on a given measure.

ratio - scale that has equal intervals and an absolute zero; one can add, subtract, multiply, and divide and one can say something is twice as much as something else, for example, a ruler.

regression effect - simply stated, this term describes the tendency of extreme scores to move toward the average on subsequent measures. Regression also means the same as correlation.

reliability - a value indicating the consistency of results of a given measure.

robust - this is a term that has been used to indicate that the underlying assumption of a test of significance can be violated with having very little effect on the accuracy of the test.

sampling - taking of a portion of a population or a universe.

score - this is a quantitative expression of a student's performance on a given measure; a score in and of itself has little meaning until it is combined and compared with a total group of scores.
simple effects - the individual groups that make up a factorial design.

source of variation - first column of an analysis of variance table which is made up of the components of variability for that particular study, the main effects, interaction, and total.

split-plot designs - see nested design.

standard score - this type of score reflects a transformation of a relatively meaningless single score into a form which expresses the score in terms of the scores of the total group, thus providing a relative meaning to the single score.

standard deviation - this is a measure of variability of a group of scores; within a normal distribution, the standard deviations (S.D.) will be a value in which $\pm$ four S.D. will encompass approximately 100% of the scores and $\pm$ one -S.D. will include approximately 68% of the scores.

statistic - a measure calculated on a sample to infer certain characteristics of the parameters of the population.

statistical significance - the probability that something is likely to occur other than by chance.

systematic variance - variability in measurements due to some known or unknown causes that effect the dependent scores to lean in one direction more than another; there are two types, biased and experimental (between group) variance.

substantive hypothesis - (sometimes called a research hypothesis) states that a relationship between two or more variables exists.

theory - a set of constructs and definitions stating the relationship among certain events or variables for the purpose of predicting and explaining the relationships between the variables.
trend analysis - is a curve fitting technique in which one tries to fit either a straight or curved line that tend to best reflect the relationship over a number of repeated measures or observations.

unbiased sampling - a sample that uses randomization procedures.

univariate - when there is one variable being predicted by another variable.

validity - simply stated, this test characteristic refers to the quality of actually measuring the behaviors which the instrument is designed to measure.

variable - something that has measurements attached to it and is under investigation; for example, 10 is a variable and the numbers may range from 0-180, sex may be a variable with numbers ranging from 0-1 or 1-2, etc.

variance - any set of measurements that has attributed to it a total amount of variance; total variance can be broken into two major components: error variance and systematic variance (see error and systematic variance); the standard deviation squared is also called variance.

INDEX

About the Authors

Dr. Isadore Newman is a licensed psychologist and a Distinguished Professor at the University of Akron where he has been teaching research courses since 1971. Specializing in mixed methodology and the general linear model, he has served on more 300 dissertation committees, authored approximately 120 articles and ten books and monographs, presented approximately 300 refereed papers, and has been the evaluator on over $10 million in grants. Dr. Newman is an Adjunct Professor at North East Ohio Universities College of Medicine and Associate Director for the Institute for Life-Span Development and Gerontology. He also has been the editor and on the editorial board of a number of journals.

Dr. Carole Newman is a Professor in the Department of Curricular and Instructional Studies at the University of Akron. Since 1990 she has taught graduate and undergraduate courses in teacher education, served on many dissertation committees and she has guided classroom teachers in their action research. She is a successful grant writer and she presents workshops on a range of professional development topics. For the past fifteen years Dr. Newman has also worked at the state and local level to design programs that support and assess entry-level teachers.

Dr. Russell Brown is a practicing Licensed Professional Clinical Counselor in the State of Ohio who has worked for over a decade to address the mental health needs of children in community, home and school settings. As the Director of Assessment for a large urban school district, Dr. Brown has had an active role in shaping the educational process of thousands of students. As an adjunct faculty member of local universities, he has taught both statistics and research methods courses and has presented at the national level on both educational and mental health issues. Dr. Brown has also authored a number of articles addressing specific methodological issues in statistics and research methods.

Dr. Sharon McNeely is an educational psychologist in the Department of Educational Leadership and Development at Northeastern Illinois University. Since 1986 she has taught courses in pre-professional and professional teacher education, educational leadership, and social work. She conducts professional development sessions to help [illegible] teaching and leadership competence, and has au[illegible] area. She has also evaluated a number of [illegible]